Duct Tape and Plastic Sheeting
A People and Nation Unprepared

Ed Walker

PublishAmerica
Baltimore

© 2005 by Ed Walker.
All rights reserved. No part of this book may be reproduced, stored in a retrieval system or transmitted in any form or by any means without the prior written permission of the publishers, except by a reviewer who may quote brief passages in a review to be printed in a newspaper, magazine or journal.

First printing

All characters appearing in this work are fictitious. Any resemblance to real persons, living or dead, is purely coincidental.

At the specific preference of the author, PublishAmerica allowed this work to remain exactly as the author intended, verbatim, without editorial input.

ISBN: 1-4241-4111-7
PUBLISHED BY PUBLISHAMERICA, LLLP
www.publishamerica.com
Baltimore

Printed in the United States of America

"A prudent man foresees the evil, and hideth himself: but the simple pass on and are punished." - Proverbs 22:3

The prudent man looks ahead at difficulties and prepares for them; while the simpleton goes blindly on and suffers the consequences

In the months leading up to the first Gulf War with Iraq in January 1991, then under the leadership of Saddam Hussein, American and Israeli officials warned that Saddam Hussein had means and a willingness to use poisonous gas against his enemies. He had demonstrated this willingness during the 1980's following an Iraq Iran war when in retribution for Kurdish support for Iran he gassed and killed thousands of Iraqi Kurds. It was postulated that Saddam Hussein would retaliate against American forces, interests in Saudi Arabia, and Israel with gas attacks utilizing Soviet built SCUD missiles. It was feared that these missiles had been armed with the capability of delivering various forms of poisonous gases including mustard and blister agents. As a precaution, gas masks were issued to Israeli men, women, and children in Israel. Some American oil workers and their families were also given gas masks in Saudi Arabia. Children and adults were advised to carry their gas masks with them and images were broadcast on American television of children carrying gas masks to school with them and drills of children donning their gas masks.

This fear of imminent attack also resulted in instructions to families to construct a "safe room" within their homes within Israel and Saudi Arabia. This was to be the shelter that would help protect the occupants from agents dispersed in a missile borne gas attack. The safe room was identified as preferably a room without an outside window, and using duct tape and plastic sheets the room was to be sealed from outside air sources by covering the inside walls, air vents, and windows. At the first threat and sound of the alarm, usually a wailing siren, the family was to escape to that room, don their gas masks, and complete the sealing of the room from the inside over the doorway entrance, using duct tape and plastic sheets.

Had there been a true gas attack using the unguided and somewhat unreliable SCUD missiles, it is highly unlikely that the duct tape and plastic sheets would have been effective. The very nature of a gas is particle dispersion in an airborne environment. Absolute protection would have required an airtight room. The gas mask would have been the major source of protection. The argument for the "extra protection" of duct tape and plastic sheets was to prevent skin exposure to blister or mustard gas. In reality, it was a story to make the general population believe that their governments could offer them some degree of security in an insecure world. It was an example of governmental public relations and spin at its very best.

It is now fifteen years later. We have seen two terrorist attacks against the World Trade Center and one against the Pentagon. We have witnessed 43 brave souls who counter attacked on an aircraft which was destined as a fourth guided bomb attack on 9-11. We have counted 168 souls lost in a homegrown terrorist attack in Oklahoma City. We have experienced massive hurricanes, from Andrew in 1991 to Katrina and Rita in 2005. Florida experienced four hurricanes back to back in 2004 and was struck by eight hurricanes in a period of fourteen months. It took four days for military aid and support to arrive in South Florida after the arrival of Hurricane Andrew which had been forecasted to strike south Florida for three days. Regardless of the pre-planning and pre-positioning, the time duration for aid to reach the victims of disaster is always measured in days rather than hours. Weeks after Katrina struck the gulf coast regions of Mississippi, Alabama and Louisiana, some coastal areas of Mississippi and Alabama had still not seen any Federal response. Evacuations of a flooded New Orleans continued ten days after the disastrous strike of hurricane Katrina. The federal government had completed a simulation of the disaster that struck New Orleans only the year before. Yet the Federal response took days for the delivery of food, water, and medical help.

During the 1990's, after hurricane Andrew struck south Florida, FEMA (Federal Emergency Management Agency) was strengthened and given the responsibility to be the U.S. government's focal point for natural disasters. They are chartered to respond to earthquakes, hurricanes, and floods. With billions of tax dollars going to FEMA every year, many subconsciously perceive FEMA as the safety net to catch them should disaster strike. Still the most immediate response to disasters has been from the American Red Cross, the Salvation Army, and other charitable agencies. They are at the sight of the disaster well in advance of the arrival of Federal and sometimes even the official State response teams. In some instances, insurance adjustors have arrived before Federal response.

Regardless of the efficiency or inefficiency of our governments to respond to disasters, it is important for us to be prepared and equipped to survive as a family for some period of time. We must be able to be self sustaining in cases of disaster until outside aid and relief can arrive.

Disaster can take on many forms; national or criminal, local or personal, and natural or medical. For example, thousands of gallons of hazardous chemicals are transported through our cities and towns by rail and truck each day. A simple train derailment could put our families in imminent danger. We should be prepared to take our families and flee to safety on an hours notice. Waiting for directions from our governments to flee or stay could result in our families being added to the total count of victims.

Chapter One

"the only thing we have to fear is fear itself—nameless, unreasoning, unjustified terror which paralyzes needed efforts to convert retreat into advance." Franklin Delano Roosevelt 1933 inaugural address

Fear

Fear mongering is defined as spreading discreditable, misrepresentative information designed to induce fear and apprehension.

There have been a number of documentaries, prophesizing mega-tsunamis, life extinction events as the earth is struck again by another asteroid of astronomical proportions, or massive tsunamis caused by underwater avalanches on the west coast or rifting of the Cascadia fault line or some other massive shift in an underwater tectonic plate. The two million acre Yellowstone National Park is a caldera for a mega-volcano. It has erupted over millennium several times. Were it to erupt today, the eruption would result in the death of millions of people and could cause a nuclear winter and another ice-age. In the geologic history of planet earth, this planet has suffered from tens of millions of strikes by asteroids and comets. Our moon was formed and held in orbit by the gravitational pull of the earth as the result of a solar collision. A portion of the new planet earth spun off into space creating our current lunar view. The planet has experienced hundreds of millions of earthquakes and hundreds of millions of volcanic eruptions. Men and women of science have tried to provide intellectual input to place these events in geologic time elements.

Science supports that in a geologic timeline for the earth that Yellowstone, will once again erupt. Historically, it has erupted at a 640,000 year cyclical rate. The last eruption was approximately 640,000 years ago. If scientist could predict with ninety-nine percent accuracy, the window is still plus or minus 6,400 years or slightly less than a hundred lifetimes for the average human. New Ice Ages will occur again on planet earth. Mega asteroids and comets will continue to strike our planet. There will be tsunamis that will strike both the western and eastern shores of the United States. Landslides will occur and volcanoes will erupt. Earthquakes occur daily and some will result in massive damage and extreme losses of life. We live on a planet which continues to change daily.

Another pronouncement is the earth is getting warmer and the polar icecaps are melting. This will flood all coastal cities around the world. In the four billion year history of our planet, there have been thousands of climatic cycles. Geologists' have studied the climatic changes in the earth and it is only the most recent one hundred thousand years that the earth's climate has had some level of stability. As recently as ten thousand years, the earth was experiencing another ice age.

When the caldera at Yellowstone National Park does erupt, or when the next large asteroid or comet does strike the earth, man can only strive to survive the aftermath. As with asteroid strikes or volcanic eruptions, disaster strikes with little warning. Millions live in earthquake zones, from New York City, Memphis Tennessee, Alaska, and California. Many in the Northwest as well as Hawaii live in the shadow of active volcanoes. Millions more live with the threat of hurricanes and the destructive forces of wind and water. Hundreds of thousands live in areas prone to flooding. Tens of thousands more live in the brush filled canyons of California or in small mountain towns and hamlets in Alaska and the Rockies surrounded by forest. Each year, fast moving brush and forest fires strike within the United States. Each year blizzards strike the plains states.

Each day through the major cities and small towns in the heartlands, railroad tanker cars carrying flammable substances and poisonous gases pass through or are switched off to rail yards. These cars may remain on the side awaiting connection to another train for a few hours or several days. A single train derailment or a leaking tanker car could endanger tens if not hundreds of thousands in a small area.

We live in a country where the infrastructure has been in place for scores of years and may prove to be our weakest link. We have only to look back to 2003 at the power outages in the Northeastern United States that plunged millions into darkness in less than five minutes. In that instance, power was off in some homes for several days while efforts were made to repair the infrastructure. Our lives and the environments we have chosen to live within cause us to accept certain risks each day. Often we as a society and nation are not prepared should a risk become a reality.

Our culture, our way of life, our arrogance, and opulence have driven many cultures and governments to hate the United States. Our efforts to share our beliefs, and convert enemies into friends will need generations and patience to bear fruit if we are to be successful. To others, those efforts appear to be a desire to force our beliefs and way of life on them. In most instances, their cultural heritage and their country is far older than our two hundred and thirty year old democracy predating us in the historical records in some cases by thousands of years. We are an impatient people desirous and expectant of immediate gratification and results. We easily bore especially as we begin to realize the cost and the time duration needed to cause effective change. For five years and at the cost of over five hundred thousand lives, our country suffered through a bloody civil war. With the current technology and commitment of the jihadist, our civil war would have been much more deadly to the general populous had the commitment of the jihadist existed in the South. For nearly one hundred years after the formal end to the civil war in the United States, the terror of the Ku Klux Klan rang in many southern and border states. There were bombings, and execution style murders, just as there are bombings and execution style murders today in the Middle East. Only because of time, technology, and different cultural belief system is the reign of terror by the jihadist considered more abhorrent.

Beyond the natural disasters that threaten us daily, there remains a threat of terror. We learned on 9-11-2001 that terror strikes without warning and so vivid are our memories of where we were and what we were doing when we first heard the news of the Tuesday morning of 9-11, we can instantly recall that day. We were incredulous at first with disbelief. Our first attempt at rationalization was that it had been a tragic plane crash. Then the second tower was struck. Still we thought the worst was over. It was then that reality began to establish that these two crashes were not coincidental or an unfortunate accident. Then a report that there had been two hijackings reported out of Boston Logan airport another in New Jersey and a fourth in Washington D.C. We knew that no airline pilot could or would intentionally fly his aircraft and passengers into any fixed object so we could not comprehend. It was only much later that we learned that the hijackings were a team effort with their own suicidal pilots having trained in American operated flight schools. The day was consumed with NEWS. We wanted to look away but we found ourselves struggling in disbelief to understand. The hijackings had not stopped at two and the day's carnage would not be known for months. We were shocked and angered that the terrorist dared to strike at our country. We had seen the attacks in the Middle-East and Europe but we never thought we could become a target. After all, this was America and in our eyes the most civilized, open, and accepting culture in the world. Why would anyone assault us? Weren't we protected by distance?

We also learned how little imagination we have. For all the action movies we see, whether it is the *Towering Inferno* or *Daylight* we do not fathom the collapse of a single tower, much less the twin towers. To us a fire in a tunnel filled with people and wrecked automobiles create a fanciful escapism movie for the end of our work week. But a real tunnel fire in a two mile tunnel in the Alps of Europe killed scores in minutes. In Boston, New York, and the Chesapeake Bay area of Virginia, thousands commute through long underwater tunnels everyday. If we can think of it, we can be certain it has been thought of by others of ill intent. The four leaders of the terrorist groups on 9-11 were well educated. We know now a lot about the nineteen terrorist who struck that day; their country of origin, how they entered the United States, how they were financed, their marital status. We know in hindsight that we should have been more attentive.

Our enemy is a group of strongly committed individuals who are committed to killing as many as possible in their death. They do not have to be foreign borne to be a terrorist. Their belief is being in the eternal glory and presence of Allah and dying in jihad assures them of that presence. Their earthly goal of immortality is their name in story and legend. What American who was old enough to remember or watched the horror of 9-11 does not now recognize the name of Mohammed Atta? Most grade school children know who Osama Bin Laden is but few can name the Secretary of State or Secretary of Defense of the United States of America.

We watched with stunned eyes the chaos wrought by terrorism and realized with anger that terrorist had succeeded in eliminating our perception of isolation, insulation, and invincibility. We were forced to recognize that governments cannot protect us and that when we leave for work in the morning, there are no assurances that we will not be a victim of terrorism at work, in school, in our travels, or at home. The attack had come from no sanctioned governmental body but from an army of mixed nationalities with a common belief that America must fall. We also learned that our values and how we cherish life is not shared by others who are willing to die for a belief or a cause. We learned that our most unimaginable fear can be imagined and carried out. Our common fault is that unless something is repeated over and over again, we often forget what we should have learned and remembered. It is now five years later. Are we still vigilant? Do we know who is entering our country?

One of the major weaknesses identified by the 9-11 Commission was communication, the same problem identified after the first bombing of the twin towers in New York in 1993. The New York City Port Authority officials had different radio frequencies than the New York Fire Department. Battalions within the New York Fire Department could not communicate with other Battalions within the NYFD. In some instances communication between first responders was best had through cellular telephones but the circuits were jammed. Congressional investigations following the debacle of response to hurricanes Katrina and Rita again acknowledged the inability to first responders to communicate. The current plan to provide enough communications bandwidth to address this problem lies with the conversion of telecasting from analog transmission to digital transmission by 2009. It is then that the United States Congress believes sufficient frequency bandwidth will be available to solve the communication issues initially identified in 1993.

Although the attacks occurred only in New York City and Washington D. C., miscommunication and rumors caused people across the nation to flock to gas stations to fill up. Some unscrupulous owners raised their prices. People paid the gouged price. For nearly two months following the attack, the sales of guns and ammunition as reported by Alcohol Tobacco and Firearms from various sporting goods stores, including chain stores such as Wal-Mart and Kmart increased significantly. In the weeks immediately following 9-11, common calibers of hand gun ammunition such 9mm, .380, and .357 Magnum were sold out nearly as fast as it was received. The terrorist attack had achieved its intent of creating uncertainty and fear in the population of the United States.

After 9-11, billions of dollars were spent on improving our systems of intelligence gathering and analysis, and improving how we as a nation respond to a terrorist attack. We centralized all the different National organizations, Customs, Immigration and Naturalization, Coast Guard, FEMA, and scores of other entities into the Department of Homeland Security. There were widely publicized drills and exercises for various what-if scenarios, nuclear, chemical, and biological.

Even after we recognized the porosity of our borders having allowed the terrorist into the country unencumbered, still, each year, hundreds of thousands of foreign nationals illegally cross the southern most border of the United States. The department of Immigrations and Custom Enforcement (ICE) estimates that they intercept only about one third of the illegal immigrants that cross. At the end of fiscal year 2005, it was reported that nearly 500,000 Mexican Nationals had been apprehended and repatriated to Mexico by the Tucson District of the Border Patrol. Another 160,000 foreign nationals (other than Mexican) had also been apprehended. Only 30,000 of that group were deported. The balance were scheduled to report for an immigration hearing and released into the United States. Some of these illegal immigrants report being driven to Interstate bus stations by United States Border Patrol Agents. It is believed that this number also represents only about one third of the 'other nationalities of interest' that cross our border with Mexico.

Most of the illegal immigrants simply seek a better life than the one they have known. In their crossing into the United States they have left behind tons of discarded trash, strewn along the one thousand nine hundred and fifty one mile border between Mexico, California, Arizona, New Mexico, and Texas. Some of the items found include backpacks, water jugs, and clothes. Other items found in this trash include Middle Eastern prayer rugs, Arabic writings, prayer books, and prayer beads. While these immigrants may also be seeking a better life as well, some of these illegal immigrants may also be seeking to harm us. Our border is very porous often only marked by a broken down barbed wire fence and a series of fence posts. Short of placing armed guards across every mile of our southern border or building an impenetrable steel wall fifteen feet high, the flow of illegal migrants will not be stemmed. Have we forgotten the numerous illegal tunnels that have been discovered connecting Mexico with the United States? The primary purpose today is alleged for smuggling drugs and people. After the erection of the Berlin Wall, there were a number of tunnels built by both the East and West Germans to facilitate escape to freedom. The flood may be stemmed by a wall, but the stream will continue.

Unfortunately, the first real test of the consolidated organizations and the first real test of how prepared America was to respond to disaster came from forecasted events that unfolded over repeated five day periods rather than a five hour period. When we were challenged by Mother Nature, we proved to America and the rest of the world that we remain vulnerable. We were not prepared for a natural disaster which was forecast in advance. Why would we expect a better response in the event of a terrorist attack or a natural disaster which occurs without warning or forecast?

Chapter Two

"Human judges can show mercy. But against the laws of nature, there is no appeal."—Arthur C. Clarke

Disaster Strikes

Hurricane Katrina struck the gulf shores of Southeastern Louisiana, Southern Mississippi, and Southern Alabama on August 29, 2005. Although rated as a Category 4 on the Safir-Simpson scale, Hurricane Katrina's storm surge was approximately thirty feet, more closely representing its earlier category five status, then in the open water of the Gulf of Mexico. On September 24, 2005, a second major hurricane struck the gulf coast of Southeastern Texas and Southwestern Louisiana. The wrath of the two storms combined resulted in the permanent relocation of tens of thousands of former residents of New Orleans alone. The devastating winds and storm surges resulted in the city of New Orleans being flooded twice. Crypts from cemeteries all across Louisiana opened. Some floated from cemeteries. Burial vaults were torn open by the water. Caskets and the remains of the dead deposited miles away, in some cases, from the deceased's supposed final resting place.

Prior to reaching landfall, both storms had proceeded North across the warm waters of the Gulf of Mexico and intensified to Category 5 strength with sustained winds greater than 155 miles per hour. The eye wall of

Hurricane Katrina passed just east of the New Orleans, Louisiana, then the eighth largest Metropolitan areas in the United States. The storm had been forecast to strike in or near New Orleans for 96 hours with an estimated intensity of a strong Category 3 or Category 4. Based on the technology and the capabilities of science to accurately forecast the exact point of landfall, they eye passed within fifty miles of a direct hit on New Orleans.

The disaster that loomed for New Orleans had been modeled in a simulation of a fictitious hurricane named Pam the previous summer. Federal, State, and local governments were all apprised of this simulation and the anticipated results. For example, it was well documented that the levee system that surround city of New Orleans was designed to withstand the storm surge of a category 3 hurricane. They were predicted to fail should a storm stronger than category 3 make landfall near the city. The model had forecast the failure of the levee system and subsequent flooding. The model had also forecast a fatality figure in thousands. Every person living in Southern Louisiana knew that someday the dreaded storm would hit, yet there seemed to be a hidden belief, perhaps a wish, of not while I am living here. The official mandatory evacuation order for New Orleans was not issued until twenty-four hours before the storm's landfall. New Orleans, a city famous for partying while the winds of a hurricane howl and famous for selling "Hurricane" cocktails was about to suffer the effects of a major hurricane. The simulation was accurate save for the estimated death toll. Hurricane Katrina's projected path and landfall had been as accurate as science could predict. The result was flooding and the abandonment of entire sections of this beautiful city built at the bend of the Mississippi River.

To have prevented the flooding, the money and effort would have to have been committed and spent over the preceding decades. Much of the city was built on reclaimed swamp land and rests below sea level. The city had flooded in 1927 and again in 1965 with hurricane Betsy. It was the lack of action following the hurricane season of 1965 that had destined the flooding brought by hurricane Katrina. If it were not the hurricane season of 2005, then a subsequent hurricane season, inevitably, New Orleans was destined for massive flooding, destined for failure of the levee system.

The orders for evacuation had begun to be issued approximately 72 hours before Katrina was to hit. Many heeded the advice and drove North and West leaving southern Mississippi, Alabama, and Louisiana. When the evacuation was designated as mandatory, thousands more fled. It was not just the poor, the old, the infirm, that remained trapped in their homes due to status and economics; it was some who refused to believe it could happen. Thousands would later be rescued from rising flood waters by boat and helicopter. No plan existed for those who did not have the transportation means or the money to heed the evacuation orders. An untold number of those left did not have televisions or radios and may never have heard there was an evacuation order. They became trapped in the small coastal towns of Mississippi, Alabama, and Louisiana, as well as the thousands left behind in the city of New Orleans. They were left to their own means, their immediate family, and prayer.

Others simply looked at more recent hurricanes and the recent low fatality rates. Four hurricanes had struck Florida in 2004 and although there had been billions of dollars in damage, few had died. Many Floridians had ignored the mandatory evacuation orders for those hurricanes, suffered the winds and rain, and appeared on News broadcast and in television commercials. The response of the Federal government, the American Red Cross, the Salvation Army and others had been timely. So insignificant the storms, they demanded only a few days of news worthiness. Hurricane Andrew had been a Category 5 when it struck south Florida in 1991. Massive wind damage, tens of billions of dollars, but few deaths had been the result. Others in the southern coastal areas and New Orleans recalled Hurricane Camille in 1969. Some viewed Camille as nature's worst and some places that survived Camille were to be leveled later by Katrina. Camille had brought wind and rain. 269 people had died, many in the flooding that had occurred later in states like Virginia and West Virginia. Camille, for all its fame and fury, had left the Gulf Coast relatively unscathed. The ability to forecast hurricanes and the relative low death rates had lulled the United States into apathy.

Although the National Hurricane Center and the National Oceanic & Atmospheric Administration repeatedly warn that the greatest danger and most deaths caused by hurricanes is from the storm surge, few seem to heed the warnings. Invariably, as each hurricane approaches the shore, news media outlets focus on the mandatory evacuations at the beach area. They repeat the NOAA warning about the risk of rip tides and deaths due to the storm surge, while filming those attempting to surf in the higher waves brought on by the advancing storm. When the tsunami struck in the Indian Ocean on December

26, 2004, there was absolute amazement and awe at the power and persistence of water. The dead and missing numbered in the hundreds of thousands. Had hurricane Katrina struck without advance warning, or no evacuations had occurred along the southern Gulf Coast, the deaths due to the thirty foot storm surge would have been proportional for the small area the storm surge struck.

With the arrival of hurricane Katrina, apathy waned. The images of the broken levees in New Orleans, of a flooded St. Bernard Parish, of a leveled city of Waveland Mississippi, of numerous Casinos weighing thousands of tons washed ashore like a glass bottle tossed in the water. It suddenly became painfully obvious that greater thought had to be given to evacuation orders and how governments must consider and plan for all their citizens. Emergency shelters had been established in New Orleans without the prepositioning of adequate emergency supplies. Contingency for resupply of food, water, and medical support was inadequate if in place. Thousands of people became trapped on islands of elevated roadways, at the Convention Center in New Orleans and the storm ravaged Superdome.

Suddenly, the streets were full of the everyday heroes who walk among us. Hundreds of small boats, manned by people from all walks of life, plowed the flooded streets of New Orleans, rescuing thousands who had been trapped in the ensuing flood when the levees gave way. The United States Coast Guard, National Guard, and United States Army rescued thousands more from atop homes. Then the stories surfaced of the medical caregivers who carried those that were trapped with them in flooded hospitals without power up flights of stairs to safety. Of the people who were enlisted to hand ventilate those patients who needed ventilation whose lives had depended on the electrically powered ventilators. The stories of those care givers who went without food so that those who were ill could have food and water. Sadly, hundreds more had drowned in the ensuing flooding of New Orleans before help arrived. Others died because of loss of power or lack of medication or equipment failures as the days awaiting help grew longer and longer.

In time, the Federal response began to arrive to help the victims of hurricane Katrina. In some of the larger areas or areas that had received the most news coverage, FEMA officials seemed to arrive sooner rather than later. Weeks after the storm had passed; some of the more rural areas or smaller towns still reported little if any response from FEMA. Eventually, buses arrived to transport those who felt they had been abandoned by the local, state, and Federal governments at the Superdome, Convention Center, and roadways surrounded by water in New Orleans. The evacuees of Katrina, who had lived for days without electricity, without running water, without a sewer system, without the convenience of communication, and in most instances, without the knowledge of what had happened to their city and their homes were bused and flown across the United States. They were welcomed into communities with open arms. Eventually, evacuees could be found in all fifty states. Families became separated, mothers from children, husbands from wives, brothers and sisters separated because of the limited space in the boat or helicopter during the rescue or the number of seats available on the bus. Nationwide efforts were begun to reunite families. Angel Flights flew family members from one city to the next to reunite children with parents. A similar effort was begun to reunite families with the pets they were forced to leave when they were rescued. In some instances, planes, loaded only with rescued pets, were flown to destinations to be reunited with their owners.

Within three weeks, the United States Army Corps of Engineers had sandbagged and temporarily repaired the failed levees. New Orleans went from 90 percent flooded to ten percent flooded. The mayor called for the repopulation of the city. The Director of FEMA had resigned and the political blame game had begun. Congressional hearings had been called for and the President was to name a commission to investigate what went wrong with the Federal response. On the horizon, a few hundred miles due south of the Florida panhandle, hurricane Rita was on the way to becoming the third strongest hurricane ever recorded.

As hurricane Rita approached the Gulf Coast nearly one month after hurricane Katrina, hurricane Rita had replaced the politicizing of what happened in New Orleans as news. The apathy, present throughout the hurricane prone regions of the United States had been replaced by either fear or a very healthy respect for the power of Mother Nature. When an order for evacuation was given in Texas, millions of people attempted to flee simultaneously. Televisions broadcast pictures of long lines of bumper to bumper traffic headed north out of Houston Texas and all points south and west toward Corpus Christi Texas. Three million people started to evacuate in front of hurricane Rita. The three and four lane wide areas of the south bound lanes remained empty while there were reports of travel times of fifteen and twenty hours to travel two hundred miles north. People drove with air conditioners in their cars off in order to prevent overheating of the engine and under the misperception that the air conditioners in today's automobiles have a significant impact on fuel economy. Pictures streamed from aerial news cameras of cars and families beside the road, cars out of gas. There were pictures and reports of people pushing their cars and staying with traffic. Gasoline stations along the escape routes had run out of gas. No prepositioning of additional gasoline to support this large flow of traffic had occurred. As the category 5 hurricane with sustained winds of 175 miles per hour aimed northerly toward Galveston Bay and Houston Texas area, suddenly the concern became that the very people who had chosen to evacuate in front of hurricane Rita could be caught on the open road with only their cars offering some minimal protection from the storm. Adding to the confusion and increasing the difficulty of this evacuation, was that many of the initial evacuees from Katrina, had been allowed to return to Louisiana. Because southwest Louisiana was also threatened by Rita, some of those who had returned were now being forced to evacuate again.

The Texas National Guard dispatched fuel trucks to refuel the stranded vehicles to allow those families who had been stranded an opportunity to continue their evacuation efforts.

When the United States began to transition to unleaded gasoline in 1974, under pressure from the Environmental Protection Agency, the automobile industry restricted the gasoline fuel tank entry on automobiles to the smaller diameter. This corresponded to the unleaded fuel nozzle size on the gasoline pump hoses rather than the larger outside diameter nozzle of the leaded gasoline service station pump. At that time, regular leaded gasoline was cheaper and catalytic converters could be damaged when leaded gasoline was burned in a car designed for unleaded gasoline. It became an industry standard that regular gasoline refueling nozzles would not fit into the refill opening for vehicles using unleaded gasoline. The same concept exists today. The refueling nozzle of diesel fuel is not only generally green in color, but the nozzle is too large to fit into the refueling port of gasoline burning vehicles today. Most military vehicles still utilize the larger refueling port opening consistent with the old "standard" gasoline refueling nozzle. It was soon realized that the Texas National Guard dispatched fuel trucks were not equipped with refueling nozzles compatible with commercial civilian vehicles. There was a delay until the refueling trucks could be reconfigured with the appropriate refueling nozzle.

Just like Katrina, hurricane Rita lost strength as it left the open waters of the warm Gulf and approached the coastline of Texas and Louisiana. Just as Katrina had hooked North and East at the very last moment sparing New Orleans a direct hit, Rita hooked North and East, sparing Houston and Galveston a direct hit. Rita made landfall on the Texas Louisiana border as a strong category 3 hurricane. First reports were of massive power outages. There was elation that perhaps the chemical and gasoline refining industries along the Texas and Louisiana coast had been spared. There was a muffled celebration that Galveston and Houston had been mostly spared. As dawn broke, winds still made it extremely dangerous for helicopter flights. The news reports came that once again, the levees had been over topped in New Orleans and New Orleans, which had recently been pumped nearly dry, was being flooded again. Just as the Mississippi coast had been on the northeast quadrant of hurricane Katrina and had suffered the devastation of both the wind from a Category 4 hurricane and the twenty five to thirty foot storm surge, southwest Louisiana shared the same fate from the Category 3 Hurricane Rita.

Unlike the FEMA response for Katrina, FEMA had prepositioned emergency supplies of food, water, ice, generators, medical teams, and medical supplies. News agency broadcast row upon row of trucks loaded with supplies waiting at San Antonio Texas to be called in by FEMA. Buses had been contracted to move evacuees as necessary. Having bought items in response to hurricane Katrina, it was not hard to divert resources and support the area with the most recent news headline, the disaster of hurricane Rita. Even with prepositioned supplies and teams, Federal response to most major areas hit by Rita was 24 to 36 hours later. Some areas in southwest Louisiana still reported no outside help had been received; weeks after hurricane Rita had passed.

Hurricane Wilma was to strike the southern tip of Florida in October 2005 as a category three having also been a category five hurricane in the western Caribbean. The storm bombarded Cancun, Mexico for nearly thirty hours with some places in the Yucatan Peninsula receiving sixty inches of rain.

When mandatory evacuations were ordered for southern Florida and the Florida Keys, approximately 80% stayed. Wilma struck the state of Florida at about 6:30 AM October 24, 2005. Six million residences and businesses lost power. Thousands of out of state electricians flocked to Florida while others continued to work to repair the damages wrought by Katrina and Rita in the Southeastern United States. News broadcast showed people pushing automobiles to gas stations when a station had gasoline and a means to retrieve it from underground storage tanks. Still others stood in lines with plastic gasoline containers hoping to obtain enough gasoline for their generators at home. Power outage estimates were forecast for three weeks for some customers.

Chapter Three

"There are two ways to live your life. One is as though nothing is a miracle the other is as though everything is a miracle." –– Albert Einstein

Status Quo

We live in a reactionary mode and not in a preparatory mode. In the event we loose electrical power during a storm, the following events may occur. We first try to ascertain that we are not alone in this event by looking out a window to determine if the power is off in the area or just our place of residence. We then try to remember where we last used the flashlight. We then lament that we should have replaced the batteries and resolve that the next time we go shopping we will buy additional batteries. If the power is off only at our home, we try to locate fuses and the fuse box or the circuit breaker panel. If the loss of electrical power is beyond our own residence and telephone service is still available, we may call the power company, only to get a recording. We will be given an estimate of when power will be restored. If it is evening, we may elect to retire early, for without electricity, our usual evening distractions of television, computer, music, and electronic games are disrupted. Usually within a few hours, electricity is restored. We return the flashlight to a place we think we will remember for the next occurrence of lost electricity. We then busy ourselves with restoring the various electrical appliances' clocks to the correct time. Unlike other countries in the world, loss of electrical power is a seldom occurrence in the United States. We will probably forget to buy additional batteries the next time we visit a store and we don't consider buying additional emergency lighting for our home.

When a powerful storm threatens homes and businesses in the United States, our 24/7 news coverage brings us up to the minute reporting of volume purchases of plywood. We also see the reports of individuals buying quantities of nonperishable food, bottled water, flashlights, and batteries. All this reaction to a weather forecast is in preparation to do battle with Mother Nature's furies. Our first plan is to hold our ground and fight Mother Nature, thus proving to ourselves and family, mankind is the master of his environment. With advanced warning through modern technology, we feel that we can prepare ourselves to survive the worst of Mother Nature. The counter point is Hurricanes Andrew, Camille, Floyd, and Katrina and their dead.

What if the do-it-yourself center is closed or the plywood and pressed board has been sold out? What is the reaction plan when the grocery store has sold out the available bottled water and batteries? Our first primal urge for survival is to avoid accepting unwarranted risk and in most instances we seek to avoid danger. We have allowed ourselves to believe our lives are less important than the material things we own. In most instances, the places we try to protect and defend are truly owned by someone else whether a landlord or mortgage company. We try to board up, stay with, and defend during Mother Nature's tempest. We try to save and protect material items often owned by someone else at the peril of our own and our families' lives.

Thanks to the constant barrage of news that tends to focus on the worst of mankind, we have seen others take advantage of voids that may occur in law enforcement. Such lawlessness in mankind may arise after a disaster caused by nature or the lawlessness of other men such as riots. We witnessed a primal regression of some to anarchy after hurricanes Katrina and Rita. We have witnessed the looting and lawlessness that has occurred after riots in major metropolitan areas in all areas of the world regardless of provocation. Looting has occurred during protests of governments, meetings of the World Bank, following major sporting events both in victory and defeat, as well as riots fueled by dozens of real or perceived inequities. We wish to believe that such lawlessness would not occur in our city or town but all men and women who share the same community have not necessarily accepted what others see as the distinction of civilization's laws. Anarchy lurks just below the surface when survival instincts or greed comes to the forefront of the mind of mankind.

The attacks of 9-11, challenged our beliefs in the rules of civilization. Suddenly the certainty of our ability to survive was challenged. Gun sales increased in the last quarter of 2001 after decreasing throughout 2000. There have always been increases in gun sales following incidents of civil unrest such as domestic terrorist attacks and domestic riots. For example, following riots in 1992 in Los Angeles, Las Vegas, Nevada as well as other large metropolitan areas in America, reported increases in hand gun sales and gun registration. All of these actions are a natural response thanks to our prehistoric ancestry. We now are faced with the need to respond to the fight-or-flight conditions thanks to the immediacy of 24/7 news broadcast which bring the scenes of disruption to the norms of civilization we expect. We are conditioned to respond. We seek to prepare to defend ourselves against all who would challenge what we believe exist in our "civilized" norm. We feel that in order to be prepared, we need a full tank of gas in case of flight, and a weapon and ammunition in case we are required to fight. We react to events that have already occurred and whether we choose fight or flight, we are reacting to history.

Clinical Psychology identifies "Acute Stress Response" as when a fearful or threatening event is perceived, humans react innately to survive: they either are ready for battle or run away (hence the term "fight-or-flight response").

No one sat down with an early Neanderthal or hominid family group 35,000 years ago and describe what they needed to do to survive day after day. Early humans survived as a result of a series of decisions made on a daily if not hourly basis revolving around what we term today as "fight-or-flight". In most instances, flight was the preferred choice. Few will find a DNA link to the Neanderthal or hominid that stayed to defend his cave against the Saber-Tooth cat or *Ursus Spelaeus* (Cave Bear). The cave bear was the largest Carnivore in Pleistocene Europe. The cave bear's history is known from an unusually great number of preserved remains found in the limestone caves of Europe where they were in direct conflict with early man. The canine tooth of this large prehistoric bear was approximately 4.5 inches long.

As a result of thousands of years of responding to stressful conditions with flight as the preferred choice, the human body developed a physiological response. As a result of survival of the fittest and fastest, this physiological response is present in all of us today. Its hallmarks are an almost instantaneous surge in heart rate, blood pressure, sweating, breathing, an increase to metabolic rates, and a tensing of muscles. Enhanced cardiac output and accelerated metabolism are essential for mobilizing fast action. The person becomes alert and attentive to the environment and the immediate surroundings.

Our lives have become more predictable and our environments safer over the last 35 millenniums. The excitement and rush resulting from an acute stress response event is now a seldom occurrence and normally reserved to rush hour traffic and the effort to avoid an accident or an unexpected challenge in our daily jobs. Some jobs are more prone to repeat of the fight-or-flight experience. First responders of law enforcement, firefighters, and emergency medical technicians are a few high stress occupations. Many have been known to talk of the adrenaline rush after a first response. Some have been known to describe themselves as "adrenaline junkies". For many of them, they must "unwind" in order to fall asleep. For most of us, after we wind down the days activities, the adrenaline rush we may have experienced normally means a sounder sleep perhaps interrupted by a dream we fail to be able to recall in the morning.

We have been conditioned to believe that as Americans, we can survive whatever nature or man throws at us. We always bounce back. We believe that our country is better after every disaster we have experienced and every threat or attack we have suffered. We are proud and resilient. We have local, state and Federal governments to rush to our aid regardless of the calamity. We live in the most powerful country in the world and we have every right to stay, protect, and defend our possessions against all comers including Mother Nature. Since the end of the nineteenth century, we have fought all wars on foreign soils. We are safe in our world and we are proud.

We are not required to be prepared. If we choose to try and weather Mother Nature's tempest we believe we can rely on our various governments and charitable organizations to be there and provide anything we may truly need. It is true that they have prepared for us. Somewhere in the last one hundred years we have come to believe that we are no longer responsible for ourselves or our own preparedness to survive. Perhaps we have been conditioned to believe that there is a safety net that will save us if we fall. If governments fail us, it is the fault of the politicians or the bureaucracies should we need food, water, or medical help either during or after the disaster has occurred and it is not there. We have forgotten that our ancestors were not the ones who stayed in the cave and defended it against the Saber-Tooth cat or cave bear. Our ancestors fled the cave and did not go back to identify the remains. We have forgotten that our ancestors prepared for lean times and poor harvest by canning and storing meat and grains in order to survive.

The Secretary of Health and Human Services recently commented that it is how the individual prepares in the one hundred hours before the disaster that will determine their chances of survival. Unfortunately, terrorist, earthquakes, tornadoes, forest fires and blizzards do not always provide a warning one hundred hours before they strike. There is praise and accolades when the response of the governments to an area in distress occurs earlier than seventy-two hours later. Simply stated, we must be responsible for ourselves and our families, including our pets. We must plan to be self sustaining for a minimum of three days regardless of our flight or fight decisions.

Chapter Four

"He that waits upon fortune is never sure of a dinner. He that lives upon hope will die fasting." —Benjamin Franklin

Survival

We must be both ready to evacuate on hours notice and prepared to stay through the storm's tempest. As a result of hurricanes Katrina and Rita, the various government organizations, and more importantly, the populous have recognized that this country, is ill prepared for massive disaster scenarios. It is only after those two events that the federal government began to evaluate scenarios such as the possibility of a flu pandemic. Things being considered are where do you house and treat perhaps millions of people infected. How do you protect the health workers? Who receives vaccine if the quantity is limited, those already infected, the old, the young? What are the options when faced with riots that may occur when vaccines are being distributed in limited quantities? That analysis will be continued and expanded to cover natural disasters such as earthquakes and hurricanes. That analysis and simulation studies can also be the foundation of the response needed for a terrorist attack when it occurs. We have demonstrated we are not prepared today.

As American citizens we have faith that our government is prepared for an attack on this country, but it is our responsibility to be prepared to protect and defend our families until the government can be mobilized to respond. It is in our innate nature to survive. Our ancestors sought to be prepared for the storms of both summer and winter. They hunted when there was game and they collected food and grain whenever possible. They recognized that most of survival was being prepared for the challenges that they were to face. The balance was pure luck.

Our means of rapid communication and notification gives us advance warning on approaching hurricanes, wild fires, blizzards and floods. With global satellites for weather forecasting and 24/7 news, we are seldom faced with leaving out the front door in the middle of the night as the flood waters lap on our back steps. Wildfires often give less warning and less time to gather our things to evacuate. In the case of a tornado, we are grateful for fifteen minutes of notification. The most dreaded event and the event with the least warning is a terrorist attack. Regardless of how much analysis, conjecture, and projection occurs before the next terrorist attack on American soil, the attack will be surprising. We will lament in hindsight we should have expected the attack and have been better prepared.

Virtually every insurance company, the U.S. Government, Harvard Medical School, and the American Red Cross either have variations of a Disaster Preparedness Plans on their respective websites or they have hotlinks available to other websites. All repeat basically four steps. First you must get informed. You must develop a plan. You must prepare for survival. The validity of your plan and how well you have prepared must be practiced, reviewed, and then revised as required. By the act of reading this, you are already executing the first step toward being prepared. You are getting informed.

Imagine an attack which damages multiple municipalities' water supplies across the nation. Imagine an attack that disrupts the electrical grids in multiple locations. Imagine for a moment an attack on the computer systems of your bank, VISA, and MasterCard. All have firewalls and security systems to block such attacks but even the Department of Defense computers have been penetrated. Now compound that by thinking of such attacks occurring simultaneously.

At the beginning of the depression era in October 1929, people made runs on the banks demanding all their money. Many banks folded and many people lost the money they believed was safely held in that bank. Banks had been speculating in the stock market, with the money entrusted to them. The amount on deposit was greater than the amount of money retained by the bank. As part of the New Deal of President Franklin Roosevelt, Congress established the Federal Deposit Insurance Corporation (FDIC) in 1933. The purpose of this independent agency was to provide some security for bank depositors that their bank deposits would be guaranteed. The FDIC initially guaranteed that depositors' money would be safe on deposit for up to $10,000. Most Americans alive today have never given a second thought about the security or the availability of their money on deposit in the bank. Before the FDIC, it was the bank company that responded when a bank was robbed. The financial collapse of many banks in the first years of the American Depression beginning in 1929, caused many Americans to lose faith in the banking industry. By establishing the FDIC, some of that faith was restored. Money on deposit was insured and the collapse of a single bank did not mean a total loss of the money to the individual depositor. Currently deposits are insured by the FDIC for deposits payable in the United States of up to $100,000.

We have grown accustomed to the convenience of Automated Teller Machines (ATM) at the local convenience store, service station, movie theatre, shopping malls, and banks across the nation. When we use automated deposits of our paycheck, we need only to venture inside our bank on rare occasions such as acquiring a loan, for transactions on certificates of deposit, or perhaps change for Saturday night's poker game. Many now use online banking and bill paying from home. As long as there is electrical power and the computer servers are up, we have few concerns. Our money is available as close as the nearest ATM. We can pay bills from the nearest computer linked to the internet. For the most part, we have become a cashless society, using our credit and debit cards for every convenience and purchases. The FDIC guarantees that our money is secure in deposit and we are encouraged to place all our money in the care and convenience of the bank. There is no guarantee however, through the FDIC or any other agency that we will always have access and be able to withdraw cash from our account on demand or when needed. We have become susceptible to failure of the electrical grid and/or the reliability and security of the bank's computer systems. We could have thousands of dollars in the security of the bank with no means to prove it or retrieve it.

Whenever we utilize a gasoline credit card or bank issued credit card at the service station, several things happen when you "swipe" the card. Via computer link, your card is verified and your credit availability is either confirmed or denied. In many instances, there is a GPS uplink that pin points the service station being used. In some instances, the authorization request utilizes the hardwired land line telephone system. When approved, we select the grade of gasoline we wish to use, and an electrically powered pump, withdraws gasoline from an underground tank. Once again we have become susceptible to a single point failure of the electrical grid. Without approval of our credit or debit card or without electrical power, the gasoline we need remains in an underground tank.

Checks remain as an option for purchasing items and paying bills. Most merchants, before they accept a check, process that check electronically to verify with our bank that there are sufficient funds to support that check amount. In some instances, that check is processed just as a debit card transaction and the merchant returns the processed check to you immediately. Once again the successful transaction is dependent on the reliability and availability of the computer system and the electrical grid.

We have only to speculate that there will be a failure in the systems that are currently used to authorize purchases of goods or services by credit and debit cards as well as checks. A devastating earthquake in the Southern California basin would interrupt much of our nation's electronic banking transactions. We must assume that terrorist organizations have already recognized such single point failures. If either event happens then until those systems are restored and the data recovered and remounted to alternative computer servers, we will once again become a society that uses cash. The duration for those systems to be down will be dependent on the severity of the attack or the magnitude of the earthquake. The time to recovery is often dependent on the resources available to recover and remount data. Loss of lives and our possible inability to communicate or notify the necessary people will increase the magnitude of disaster and the duration to recovery. The plastic cards we currently use, our paper checks, even travelers checks, will only have limited acceptance. We will suddenly operate as most of the world operates today, a cash only society.

By choosing to live in a major metropolitan area, whether it is a city of eight million or two hundred fifty thousand, we are a more desirable target for terrorism. Terrorism's goal is to inflict fear. An event resulting in mass casualties creates a condition of insecurity in all other sectors of the country. Fear is a weapon that has been used for centuries. Attila the Hun often decapitated the vanquished in order to make other tribes fear his wrath and his army's ferociousness. So feared was he that Rome paid tribute to Attila and to keep his armies away from Rome. Genghis Kahn terrorized the tribes of China and united them. Germany terrorized London in World War II by launching V2 rockets indiscriminately at London. The dropping of the atomic bombs in Japan created sufficient terror in Japan, that Japan capitulated to an unconditional surrender to end World War II in the Pacific. Terror is intended for the masses. A hundred lives lost in a terrorist attack in a small town in Nebraska would remain news only a few days. A hundred lives lost in a single subway attack in New York City or San Francisco would inflict fear in millions of people who use those systems daily, not only in the attacked city but other cities with a populous dependent on mass transit. A terrorist attack on the commuter train system in the Northeastern United States could paralyze the financial district and have worldwide economic repercussions. Many would simply stay home and avoid utilizing any mass transit system.

A large scale terrorist attack which inflicts death and fear in multiple communities would strain our infrastructure and our system of law and order. Such an attack would challenge many to maintain what we now deem as civilized behavior. We are a nation of laws but also a nation with a stratified populous with varying degrees of comfort, wealth, and security. A large segment of society lacks in the same comfort, wealth, and security. Chaos and a collapse of the infrastructure of the country could result in the collapse of law and order. Depending on the magnitude and the duration until sufficient forces could be mustered to restore the rule of law, it could take days or weeks for law and order to be reestablished. The effort to reestablish law and order could require the formal imposition of martial law and suppression of some of the guarantees provided by the Constitution of the United States of America.

Technically Martial Law has only been imposed twice in the history of the United States but remains an option for the President of the United States. There have been instances where the National Guard have been federalized and activated by the President on U.S. soil such as integration of the school system in Little Rock, Arkansas in the fifties and during the civil rights and anti-war movement of the sixties and seventies. The activation of the "militia" is pursuant under Article 1 Section 8 of the U.S. Constitution. It is waiver of the writ of *habeus corpus* which is guaranteed under Article 1 Section 9 and activation of the "militia" that has precedent in establishing the criterion for martial law.

It is the very possibility of such an attack, an attack resulting in anarchy that should motivate us to be able to be self reliant for several days. We must be prepared to load an Emergency Preparedness Kit, and immediately flee to a small town in Nebraska or Montana. By preparing for the worst case scenario, we will be ready to handle any eventuality whether it is hurricanes, floods, wild fires, earthquakes, or a terrorist attack.

Before the Twentieth century, most of America was rural. People had learned to live with hardship and in time of crisis were dependent upon their families and their neighbors. They also knew that their relief agency and outside support would come from their church or local community groups. When a barn or home burned or was damaged or destroyed by storms, it was neighbors helping neighbors who rebuilt the home or barn. It was neighbors who provided shelter for those whose homes were lost. Neighbors shared what little food they had. Families gathered to preserve the harvest through canning or curing in order to survive the winter. There was no television, radio or telephones. FEMA did not exist. Word of disasters or need did not reach around the world in seconds. It was communities that supported the needs of all whenever one or all found themselves in need. If a tornado struck the town, it was the people of that town and the surrounding area that picked up the pieces and rebuilt the town without state or Federal aid.

Our lives today in the United States are quite different than how most have to live around the world. Most Americans live in the comfort of centrally heated and centrally cooled homes. We have the convenience of electricity. As a result we have the means to keep and store food safely for weeks and months. We are not required to go to the market each day to purchase the food we will eat that night. We have televisions, telephones, and computers. If we own or rent a home, we usually have washers and dryers. Our homes have multiple rooms, indoor plumbing, and typically more bedrooms than occupants living in that space. We have family rooms where occasionally all the family will meet. We have a guest bedroom but seldom do we have guests visit. It is typical for a family to have two cars. In the minds and eyes of most people in most countries we live an opulent life. We carry insurance and warranties on everything we own including ourselves. We may wave at our neighbor down the hall or down the street but seldom do we know their name. We are a self consumed society looking away in most cases from the homeless man on the corner. After all, didn't the government create a 'safety net'? Our thought is that they must not want help. If they did, they need only to ask the government, but they should not intrude on our lives. We donate food to the food bank at Thanksgiving and Christmas to make ourselves feel like we are better people and that we really do care about our fellow man. What we don't acknowledge is that hundreds of thousand of Americans depend on the community food banks for their daily meals. When disaster strikes, we have developed into a society that believes someone will come to our rescue. If it is a personal disaster, automobile accident, fire, health issue, or the refrigerator breaks down, we paid the insurance premiums; someone will take on the responsibility. If it is beyond us, and beyond our insurance coverage, someone else will come. It may be the American Red Cross, Salvation Army, our church, our State or Federal government; someone will come to remove the burden of self survival.

We have become so reliant on electricity, we take it for granted. Without it, our entire distribution system of goods and services, including food and water are interrupted. Without electricity, our ten days of food supply within a city is reduced by a third. One third will rot or decay within five days without refrigeration. It takes electrical current to drive the pump motors to retrieve the gasoline or diesel fuels from the underground storage tanks at the service station. There is a slogan a trucking firm uses; 'if you go it, it came by truck'. Without electricity, the truck does not receive the fuel to move it to the store.

Much of our electrical infrastructure dates to the early and mid twentieth century. On August 14, 2003, 50 million people in Northeastern United States and Southeastern Canada were subjected to a failure of the electrical grid. Areas affected included Ohio, Michigan, Pennsylvania, New York, Vermont, Massachusetts, Connecticut, New Jersey and the Canadian Province of Ontario. The first failure occurred in Ohio at approximately 4:08 pm EST and the sequence of events in the chain reaction was complete about 4:13 pm EST. Power was not restored in some areas for four days. This failure was traced to a single point failure on the electrical grid in Ohio. After that failure, a U.S.-Canada Power System Task Force was established. In their final report published in April 2004, they presented 46 recommendations to help prevent recurrence. They identified the single most important of these recommendation was for the U.S. Congress to enact the reliability provisions in H.R. 6 and S. 2095 (otherwise known as the Energy Bill). The Task Force recommended that compliance with reliability standards be mandatory and enforceable. The Energy Bill was finally passed on July 29, 2005 and was signed into law in August 8, 2005. It is now public law 109-058.

Just as most regulatory laws, the requirements exist, but it is only after failure, the investigation, and possible fine of a company for not implementing or upgrading their area of responsibility for the electrical grid that the new regulations will be implemented. Until the provisions of the new law and the recommendations of the Task Force are implemented we remain exposed to the possibility of inadvertent human or failed hardware caused blackouts. Should we fall victim through a natural disaster or terrorist caused interruption to multiple electrical grids, the resources required would over extend our current capabilities and many areas could be without power for weeks. There are a limited number of resources, tools equipment, and manpower capable of reinstalling 100,000 volt power lines or erecting aluminum power transmission towers. As evidenced by the power outage of August 2003, the electrical grid of Canada and the United States are interwoven.

As heads of households, it falls to us to adopt the Boy Scout motto of 'Be Prepared'. We should have learned from the debacles of hurricanes Katrina and Rita that our nation is not prepared for mass evacuation. We are not prepared to respond to a community under attack with resources needed for immediate survival. We saw that with prepositioned supplies of food, water, and ice, it still took FEMA 24 to 36 hours to bring supplies to Southeastern Texas after hurricane Rita. Weeks later, some smaller towns still were without power and had not received Federal aid. In the event a community's water supply is contaminated, there is not enough bottled water already in place on the grocery shelves in the community until FEMA can arrive. A simultaneous assault on the electrical grid would interrupt the distribution system in that area by interrupting the ability to refuel vehicles. Such a simultaneous assault would also interrupt our communications, our ability to keep foods refrigerated, and our ability to purchase in a cashless society.

During the aftermath of hurricane Katrina, we saw the very beginning of the breakdown of society. We witnessed a police force without the ability to communicate. Electrical power was disrupted, radio communications failed as batteries depleted, cell phone communication also failed. We witnessed through the power of instant news 24 hours a day, a city start to revert to anarchy. Who would not loot to keep their family fed? Our natural instinct is to survive.

It is not hard to reach anarchy in the scenario given in the previous paragraphs. What would happen if we no longer believed our water supply was safe? Even the rumors of such a failure would have to be dispelled. Couple that rumor with the failure of the electrical grid. Compound that with more than one disaster in multiple cities. The ability of our support resources such as FEMA, the American Red Cross, and the Salvation Army are incapable of supporting everyone in need simultaneously. That is assuming that they have the means to even reach the affected communities. Living without our daily feed of twenty-four/seven news coverage, we may never know that the reason no one has come is because they cannot reach us or do not know we are affected also. Suddenly our lack of self reliance and always counting on someone else being there to respond to help us survive could be fatal.

Chapter Five

"In preparing for battle I have always found that plans are useless, but planning is indispensable." – **Dwight D. Eisenhower**

Preparation

Shelter can take many forms from the most basic depression or cave in the side of a hill to an elaborate underground bunker system built during the cold war. Although there is less focus on them today, many of those are still found in major metropolitan areas. Some are still maintained and remain stocked with supplies sufficient to remain self sustaining for a specified capacity for up to six months. If you are concerned about the possibility of a nuclear attack, plans for building a personal fallout shelter can be obtained free from FEMA. There are also a number of options presented on various websites from building a lean-to in your basement to buying and burying a pre-fabricated tubular shelter. Shelter can be had in a basement or even a home. Whether you decide in the future to use your home as shelter against the storm or you seek shelter at a facility in your area designated as a shelter, you are responsible for having enough food and water to sustain you and your family for a minimum of seventy-two hours. What was witnessed in the migration of people to shelters in New Orleans during hurricane Katrina was a lack of preparedness on individuals and/or their inability to bring their emergency supplies with them as the result of the flooding. What also became evident is that some events occur that will result in large masses of people congregating who need food and water immediately. Such events as the flash flooding in

New Orleans, a tsunami in Oregon or Washington State, or earthquakes in California or Missouri does not alter our requirement to prepare, protect, and defend our families for three days without outside aid and resupply.

Thousands have chosen to ignore evacuation warnings in the past, instead hoping and believing that the disaster will not happen or be as bad as predicted. With the proper preparation, staying can be a viable option. If your home is built high on a hill and the flood plain is below you or your home has an underground storm shelter or an internal concrete and steel reinforced safe room, staying to weather the storm may be an option. However, you must be prepared to be isolated without power, communication, and expect no outside resupply or help to arrive for several days if not weeks.

Historically, people in hurricane evacuation areas have been advised that they should be prepared to survive ninety-six hours without the arrival of aid. Those four days seems to be the expected response time for FEMA support to large populated areas after being struck by a hurricane. On the other hand, the American Red Cross indicates preparation for self sustaining should be at least seventy-two hours. Those times are all based on after the storm has passed. Hurricanes have been known to pass quickly, while others, especially the more powerful storms, may take several hours or even days to pass. For example, a category two or three hurricane may have sustained hurricane force winds extending up to fifty miles from the center. Gale force winds may extend another fifty miles beyond the hurricane force winds. In this example, the diameter of the storm containing gale force or hurricane force winds is simply two hundred miles. At a forward speed of ten miles per hour, your shelter can expect to be buffeted with gale and/or hurricane force winds for approximately twenty hours. It is only then that the clock starts as to the response time of FEMA and charitable organizations in the wake of a disaster. Of course, the timelines published assume that accessibility remains intact. Some areas of southern Mississippi and Alabama coastal areas as well as small towns in western Louisiana and southeastern Texas experienced weeks without the arrival or support of FEMA assistance. Roads and bridges had washed out. It is as easy to prepare for four days to live autonomously as it is for six or eight days. The most difficult thing is recognizing that disaster can strike anywhere and anytime.

Where do I start?

The short answer is you have. Become informed about the hazards that exist around you and in your community. Become informed about places and facilities within your local area designated as shelters. Much of this information gathering can be started by contacting your local American Red Cross Chapter. Be aware of your surroundings and look for evacuation plans posted where you work or in places you visit. All hotels, motels, restaurants and businesses are required to clearly post evacuation routes. In every elevator, there is a clear warning that they are not to be used in case of fire. Some automatically are disabled. Learning where the staircase is while trying to crawl down a smoked filled hallway is not planning to survive. It is simply blind luck if you do survive. Review your home for hazards and formulate an evacuation plan for your family. Your home is the most likely location for an injury accident to occur. It is very important that children react to a plan in times of extreme stress such as an earthquake, tornado, or fire. The optimal word is react. Drawing out an evacuation route on a piece of paper for your home and explaining it to a child will not be remembered by a child with a house filled with smoke or flame. Evacuations must be practiced and alternative routes and conditions must also be planned and practiced. It should be a game and second nature to young children.

Of all the possible disasters that can strike, the highest probability of disaster is a personal disaster in your home from fire. The forgotten candle, the space heater improperly positioned or knocked over, a stove burner left on, a child playing with matches, or with the cigarette lighter left on the coffee table. It could be caused by the careless discarding of still smoldering smoking materials or that one last cigarette before you fall asleep, possibly for the final time, in bed that night. It does not take a tornadic wind to blow a dead branch or dying tree into your house.

A light breeze coupled with time or simply time and gravity will also cause a branch or tree to fall. All things being equal, that limb or tree will fall on your house, your car, or your garage. The spark from a snapped electrical wire against your house offers the potential for multiple disasters from electrocution to fire. A broken or leaking gas line has the potential of lethality if undetected, an explosion if ignited in a highly concentrated environment, and fire. A faulty central heater or gas water heater can fill your home with carbon monoxide gas, a colorless and odorless gas which reduces your body's ability to absorb oxygen until death occurs. By purchasing and installing battery powered smoke and carbon monoxide detectors, sufficient warning may be available to save your family and yourself from some of these scenarios. As with all things electrical, the batteries must be periodically replaced and the detectors must be tested to insure they are operable. Simple awareness of your surroundings and your environment may alert you of hazards. That dying tree or rotting limb may be best removed before significant damage can be done as a result of time and gravity.

Beyond looking internal to ourselves and internal and external about our home for possible disasters, we must consider the possibility of outside threats. What happens locally? Am I in the tornado zone? Does the area experience ice storms that can take down electrical power for days or weeks? Do I live in a flood plain or in an area where access is limited and can be cutoff as a result of flooding, mudslide or rock slide? Am I inland from the immediate affected area of hurricanes but in an area which is prone to lose electrical power for long periods of time as a result of hurricanes? Is my home in the valley shadowed by an earthen dam? Are there special needs in my family such as a member needs dialysis, special medical treatment at a hospital, or is there a need to keep medications such as insulin refrigerated? Whether it is flooding or tornadoes that cause us to be isolated or cut off, or the collapse of the electrical grid from age and maintenance, to the more sinister terrorist attack, we must be prepared to keep our families alive with enough supplies for a minimum of three days. The more we understand about the possible threats both natural and man made, the better we can prepare our homes and families to survive.

What do I need?

Your needs and your family's needs are for whatever it takes to keep you and your family alive for three to five days. Hopefully, you can establish and maintain shelter in your own home or some other personal shelter such as a storm or fallout shelter. If you must seek shelter in a public shelter, you must still be prepared to have enough food, water and other supplies to sustain yourself and family for a minimum of three days.

What you will need depends on the answers to the previous questions. If you have a family member that needs dialysis, you may need to evacuate if at all possible to an area that has a hospital or medical center that can perform dialysis. It is possible, but not probable, that a public shelter may be equipped to support this kind of need. If there are other medical needs such as chemotherapy or radiation treatments on a scheduled basis, you also need to try and evacuate if possible. Realize the goal of relief in seventy-two to ninety-six hours is simply a goal. Larger, more news worthy areas, will always receive the relief earliest. Depending on the magnitude of the disaster facing the relief agencies and the government organizations, relief could be a week, weeks, or more away. If there is medication that requires refrigeration, you may wish to buy a small refrigerator and a portable electrical generator or perhaps a refrigerator which operates using propane. Of note, portable electric generators should only be operated in an open area. There have been fatalities from people running the generator inside their homes as a result of carbon monoxide poisoning.

You need to prepare early. Unlike hurricanes, most disasters happen within a few hours or even minutes, rather than being forecast and projected for several days. A mudslide or rockslide takes only an instant although they may be preceded several hours or days by rain. Earthquakes may last a few seconds to a minute or more. They are not forecast or predictable. Although a volcano may rumble and spew, the actual eruption, magnitude, or direction of the eruption cannot always be accurately forecast. Of the fifty-seven people killed at the eruption of Mount St. Helen in May 1980, most were outside the area designated by the USGS as the red zone and not in the expected path that the eruption took. There is elation when tornado warnings arrive fifteen minutes before the storm and allow people to seek immediate shelter. But in the open prairie and on the farms that occupy much of tornado alley, it is the sound of the wind or a visual sighting that produces the warning for the family to seek shelter. An ice storm can be forecast but the severity and the amount of damage is only experienced. Terrorist attacks occur without warning and can take any number of forms, from explosive and localized to coordinated simultaneous attacks across a wide area of targets.

When preparing stores of emergency supplies, keep in mind that the preparation must assume that you will need to gather your stores and leave your home. It is easier to load out when can goods and other items are containerized and centralized. Assume that when the time comes to load, you are alone and need to load out immediately. Pack in small and light enough amounts that a single person could execute the loading operation. If more than one person is there to perform the task, the task can be completed in half the time. Containerization should be done before the emergency occurs. With today's uncertainties, you may wish to be able to gather and evacuate your family, with your emergency stores in minutes not hours. There may be no warning as to the expectation of a disaster such as a notice of several hours in advance of a hurricane but rather the sickening few seconds when a major earthquake or terrorist attack occurs. By keeping as many of the following items on hand as needed and keeping those items centrally collocated and containerized you will be prepared to support your family for a number of days while relief efforts are mobilized. If conditions are such that you must evacuate, your families immediate needs can be met by quickly loading prepositioned stores in the vehicle being used to evacuate.

Your ability to retrieve cash from your bank account or credit card through ATMs may be lost. Places that do remain open to provide shelter or food after a disaster may want cash customers only. It is possible that some gasoline stations will be open with pumps powered by portable electric generators. Gasoline may be more expensive and those service stations may seek cash customers only as well. It is well to plan expenses at $200 per day for food, lodging, and fuel if available. It is not unreasonable to have one to two thousand dollars in cash in reserve in your home. Larger families may elect to have larger amounts.

For a family living paycheck to paycheck, of course the thought of such an amount in personal reserve at home is exorbitant. The natural question is where does he expect me to come up with two thousand dollars? We all have pocket change and typically throw it on the dresser at night. Some eat lunch in the cafeteria or local deli every day. By collecting the pocket change and simply taking your lunch one to two days a week or even skipping lunch one day, the amount and rate of savings can be surprising. Provided the "piggy bank" is not raided for such things as money for sodas, beer, or the children's school lunch. Collecting hundreds of dollars in three to five months is not out of reach. Cash should be secured in a lock box or safe in the home, in smaller denominations of bills, tens, twenties, and fifties. Although travelers checks are convenient and more secure, they may not be acceptable in a wide spread event such as a terrorist attack in multiple cities or locations. This cash reserve is not for Christmas, birthday, vacation, cruise, or other "emergency" expenses. This reserve is for the day you hope never occurs.

The following is a compilation of items and recommendations from various organizations including FEMA, the American Red Cross, as well as Harvard Medical School. All three begin their list with the first item on this list as well.

Bottled Water: It is recommended that 1 gallon of water per person per day be stored. One (1) case of half liter bottles (16.5 oz.) of water is approximately two and a half gallons. Approximately five (5) cases would be recommended to sustain a family of four for three days.

Optional: Five (5) gallons of potable (drinkable) water in plastic jugs or containers designed for water storage. Thoroughly rinsed one gallon milk jugs are excellent for this purpose whereas, a large container which holds three or more gallons is usually too heavy and too bulky for all family members to handle conveniently. This water could be used for drinking but is primarily in place to wash hands, dishes, faces, and etc. On a three day weekend camping trip with two adults and two adolescent children, five gallons of water will easily be consumed in washing dishes and utensils as well as maintaining some degree of personal hygiene.

Non perishable canned goods with a mechanical can opener: Some stores sell can goods in case lots. Other conveniences are cans with pull tabs for quick opening. The food should not require any cooking and should be able to be eaten from the can cold. Ready to eat soups, pork and beans, tuna fish, other canned meats, and most canned vegetables such as beans, asparagus, and potatoes fulfill this category. It is good to have a variety especially with younger children. Ravioli as well as other pasta type foods can be eaten cold. The labels may have such phrases as 'ready to eat', 'heat and serve', or 'pre-cooked'.

Baby Food and Pre-measured and pre-mixed baby Formula: Without electrical power, the conveniences offered as a result of refrigeration suddenly become noticeable. The ability to refrigerate and keep partially eaten jars of baby food is no longer an option. In cases where there are children of the age that require baby food or pre-measured formula, sufficient quantities need to be purchased to accommodate the lack of refrigeration.

Disposable Diapers: This is a case where it is better to have stockpiled this item in excess rather than be under prepared. The age of the child will determine the quantity as well as the type of diapers that need to be held in reserve.

Anti-bacterial products such as baby wipes, liquid hand soap, dish soap, and hand lotions: In the event of a disaster, either natural or man caused, medical aid, when available, may be focused on treating those victims with life threatening injuries. One of the best ways to limit the spread of germs and to keep ourselves and our family healthy is by utilizing some of the many products identified as anti-bacterial. Some studies have even indicated that the use of plain soap is just as effective as many of the anti-bacterial products. Regardless of the designation, hands need to be clean both before and after we break for Mother Nature, as well as before and after we either prepare or consume food. The luxury and convenience of indoor plumbing, the ability to wash our hands using water flowing in the house, the convenience of a shower or bath could be lost for days if not weeks.

Eating utensils (plastic or disposable): Have a sufficient number of paper plates and bowls along with plastic eating utensils. The plastic eating utensils may be washed and reused if necessary.

Trash bags: There should be multiple sizes of trash bags available. The larger bags should be used to collect and retain garbage. Large trash bags can also be used as a water barrier to keep items such as clothing, paper products, and documents from getting wet. It is better to have the thicker heavy duty garbage bags. Trash may need to be retained in the proximity of the family for some period of time. Smaller "kitchen" trash bags should also be available. They can be used to collect human waste in the event that water and sewer service is interrupted. Again, the heavier ply plastic bags should be preferred because of the potential uses.

Toilet: There are a number of possibilities for toilets. The most desired is that the sewer and water system, even if the water is no longer considered drinkable, is not interrupted. Ideally you have been able to remain in your home, possibly without potable water and electrical power. An alternative is the portable chemical toilet. These are relatively inexpensive but usually require special types of toilet tissue. It is possible to build a portable toilet using a little plywood, a toilet seat, and a five gallon paint bucket. A five gallon paint bucket is approximately 14 ½ inches tall, approximately the same height as a standard commode. By building a three sided (open back) plywood box 20 x 20 x 15 and mounting another piece of plywood on top with an opening and a standard toilet seat, the five gallon bucket lined with a plastic 'kitchen" garbage bag will replace the toilet bowl. The "paint bucket" can be removed from the back and the garbage bag secured and replaced. For those that are not handy with tools or do not possess the means of building a plywood port-a-potty, some sporting goods outfitters sell toilet seats that snap onto a five gallon bucket. Cost is generally less than fifteen dollars.

Human waste is a bio-hazard and must be disposed of appropriately. By building this portable toilet with removable top and hinged sides, or purchasing a bucket lid toilet seat, the amount of space required for storage is minimal.

Toilet paper, Paper towels, and Sanitary products: This is an absolute must. With two adults and two adolescents, one roll of paper towels and two rolls of toilet paper are normally sufficient for a four day weekend camping adventure.

Sleeping bags and/or blankets: When preparing for disaster, it is important to plan for conditions when there is no heat. Most sleeping bags are thermally rated, some for comfort in milder temperatures while others rated for sub-zero comfort. It is best to have sleeping bags for the predominant conditions which are better than sub-zero normally, and augment the sleeping bag by lining the interior with blankets or covering with quilts if the conditions are below the rating of the sleeping bag.

Toiletries: Soap, toothbrush, toothpaste, hair brush, mouthwash

First Aid Kit: A First Aid Kit can either be purchased or built from supplies available at most pharmacies. Regardless of the path, a First Aid Kit is a must. Most injuries can be treated in the home initially, and in the case of disaster, medical treatment by either Emergency Medical Technician or Professional Medical Staff may be hours or days away. The Harvard Medical School has identified items that should be in an Emergency First Aid Kit and I have included them here for convenience in preparing a personal or family First Aid Kit.

Wound care

One roll of absorbent cotton
Gauze pads (4 inch square)
Adhesive tape (1 inch and narrower)
Adhesive bandages in various sizes
Butterfly bandages
Wound cleansers (soap, gels, or wipes)

Medications

Analgesic, such as acetaminophen (such as the trade name Tylenol or generic brand) or ibuprofen (such as the trade name Advil or generic brand) for both adult and child or Aspirin (for adult use only).
Antihistamine for allergic reactions
Antiseptic ointment or cream (such as bacitracin or triple antibiotic ointment)
Calamine lotion or hydrocortisone cream (1%)
Activated charcoal for inadvertent overdoses
Saline eye drops
Antacid for stomach upset
Antidiarrheal medication
Oral glucose preparation for low blood sugar

Other Supplies

Ace bandages
Cold/hot packs (some chemically activated packs are available which act as either a cold or hot pack without refrigeration)
Cotton swabs
Flashlight

Scissors and safety pins
Surgical gloves (disposable)
Thermometer
Tweezers

First Aid Manual

You may find some First Aid Kits will have many of these items but most will not have all of them. It may be more convenient to purchase a well equipped First Aid Kit, place its contents together with previously listed into a fishing tackle box. It is best to purchase a plastic, water tight and floatable tackle box. Clearly mark the box with "First Aid Kit" and make sure it is conveniently located. If the First Aid Kit is used in events other than a planned for disaster, ensure that used items are replaced as soon as possible.

Contact the local chapter of the American Red Cross and take a course in CPR and a basic First Aid course. You may be the only person that can save the life of one of your family members.

Medications: Have a three day supply (minimum) of all daily medications. As discussed previously, 72 to 96 hour response time is a goal and depending on the magnitude of the disaster, additional time may be required to reach your specific location. Once rescued, additional time may be necessary to locate and refill any prescriptions. Most doctors will authorize prescriptions for ninety days and refills can usually be obtained seven days before the end of the medication prescription.

Medical Information: A list of names and phone numbers for all the family's primary care and any specialist care physicians. A list of all medical conditions, any chronic conditions, and allergies if any. If you or your family are not able to respond when found, this information could be life saving for you or a member of your family. This information could be kept in a water proof package and inside the First Aid Kit.

Radio: Battery powered or hand generator, this is a must have item. Use an adhesive tape and identify the frequencies of the emergency broadcast and/or news channel on the radio for both AM and FM frequencies. While not as generally available, some emergency band and weather radios do have a hand cranked generator as a power source. These radios either do not require batteries or those using batteries can have the batteries recharged without removal.

Flashlight: Multiple battery powered flashlights are a must. When stored, batteries should be removed from the flashlights. Plastic, water proof flashlights are preferred. When practical, the same battery size for both the flashlight and radio are preferred.

Batteries: Multiple sets of batteries should be stored and available.

Masking tape or Duct Tape: It is said there are a thousand and one uses for duct tape. In the event of a disaster, you may discover one thousand and two. The ingenious use of duct tape on the failed Apollo 13 mission was a major contributor to the safe recovery of all three astronauts.

Plastic Sheeting or Plastic Tarp: The primary purpose is to provide shelter and covering. In the event of a storm and rain leaks through the roof, the tarp may be used either to cover the roof or initially provide shelter and cover for the family during the height of the storm.

Tools: Simple tools, a knife, pliers, can opener as a minimum. A quality multi-purpose tool can fulfill many of the minimum requirements and may help for the unexpected need. Two of the better known manufacturers are Leatherman and Gerber.

Pets: Pets are additional members of the family. Have plenty of food, water, any medications, history of shots and vaccinations, collars with tags (when applicable), leashes, and cages or carriers as applicable. They will also experience stress in a disaster situation.

Dust Masks: Especially in areas at risk for volcanic activity. Dust masks are also recommended during dust storms or high winds when loose particles of house insulation or damaged wallboard may be present in the air.

Documents and Financial Records: Insurance papers, identification papers such as birth certificates, driver's license, passport, marriage license should all be stored in a water proof package and be locked in a home safe or fireproof box. A list of any bank accounts and their numbers, as well as any investment accounts numbers, mortgages, and loans should also be kept in a water proof package and locked in a home safe or fireproof box. In case of evacuation and the destination does have banks available and functioning, United States Saving Bonds can be cashed at some percentage of their face value depending on the maturity date of the bond. In case of disaster, all these records and documents should be easily retrieved and be easily transportable with your disaster kit. Make a DVD recording of all your household possessions. In the event of a disaster, that recording will become invaluable for insurance settlement purposes.

Contact Information: Phone numbers and addresses of friends and family members. Emergency phone numbers such as the American Red Cross, Salvation Army, as well as state, local, and federal numbers.

Cash: As previously discussed, in the event of loss of power or loss of computer systems that support the banking industry, the plastic credit and ATM cards will be useless. Smaller denominations of cash will be preferred by vendors who have items you may have forgotten or need.

Cellular Telephone: Depending on the nature of the disaster, cellular telephones may be more reliable than the traditional land line. In order to charge the cell phone battery, a charger not dependent on the electrical grid, such as a car charger or some other battery powered recharging device should be available.

Family Entertainment: Family games such as Monopoly, Scrabble, or puzzles should also be stored. Such activities will reduce the natural stress that can and will occur in the event of a disaster. These will also help reduce the level of boredom while you await rescue or conditions and environment normalize.

Optional: There are a variety of devices manufactured for treatment and purification of water that can be purchased at most outdoor and sporting goods stores. Purification can occur by using the simple iodized pills that can be added by the canteen to utilizing cartridge systems that allow for purification of gallons of water on a single cartridge. Water is guaranteed to be 99.9% free of waterborne cysts and protozoa and 99.99% effective against water-borne bacteria and viruses at a minimum by most of the cartridge purification systems. These systems vary in size and cost. Some systems are manufactured small enough to be carried in a back pack which can purify one quart of water per minute.

Now I am prepared so now what

You may have stockpiled supplies, organized everything, know exactly how long it will take you to load everything in your vehicle and be ready to leave or prepared to stay depending on the situation. However, this is not the end or completion of preparation.

How prepared are you? Have you discussed this with your family? Have they been a part of obtaining, preparing, and storing the disaster supplies? Do they understand why all this is being done? After all, you and your family may become separated either due to the event or as a result of coordinated evacuation when disaster strikes. Your family's knowledge that you have planned for this day will help to reduce some of the initial stresses and fears. Have you and members of your family identified some person out of the immediate area who will be your central contact point? Do all members of the family have the name and number of this contact point? What if some members of the family are injured and separated while other members are evacuated? Having a centralized contact point may help to reunite the family in a timely manner or at least provide some reassurance to the well being of your family. Your contact should know your escape plans, any meeting points, and possible destinations. Initially phone lines may be congested or being used by first responders and inhibit your ability to get through. How ready are you and how ready is your family for the unthinkable?

Make a plan that involves the whole family. Everyone should understand the conditions that may require you to either stay put and survive in adverse conditions or load your disaster kits and flee. This can be explained to older children without instilling panic or paranoia. We live in a factual world surrounded by news broadcast twenty four – seven. Just as we are bombarded by news everyday, our children are not insulated nor isolated from world events and the reality of world events. With younger children it can become more of a game.

Practice an evacuation without the realism of stress. Take some of the essentials from your disaster kit and take your family camping over a weekend in a National

Park. (Read the next chapter before just picking up and leaving with your supplies.) It can familiarize everyone with life without television, laptop, and computer games. Camping is not staying in the local motel at the edge of the Grand Canyon or on the border of Yellowstone National Park. Others argue that camping is roughing it in the RV or 5th wheel trailer with a small screened television and a satellite is still camping. Take a tent. Sleeping on the ground may convince you that you would like to have cots in your disaster preparedness kit. It can also make you aware of items that may not have been previously identified that your family would like to include.

Chapter Six

"We must use time as a tool, not as a crutch." —John Fitzgerald Kennedy

Drill

Although the items identified in the previous chapter are consistent across several organizations, they are the minimum items needed in your disaster preparedness kit. The basics that man requires in order of importance is water, food, and shelter; all covered in the previous chapters.

A camping experience will identify additional items of convenience that have not been previously addressed.

Cooking Utensils: Granted, canned nonperishable foods can be eaten straight from the can and cold. Archeological records show that man survived thousands of years before he began to use fire for heat or cooking. When we are not in a survival situation, we prefer to have soups heated, canned meats warmed, and baked beans steaming. Although this could be done by heating the cans over or in an open fire, pots and pans are a lot more convenient. It is a lot easier to use the handle on a pot or skillet than reaching into a fire or on a grill top and grasping a warm can of beans. Have a small "starter set" of cooking pans; a 1 and 2 qt sauce pan, a large and small skillet are really the basics. Basic cooking utensils should include a large spoon, tongs, spatula, and large cooking fork. A set of oven mittens are also a must have.

Tools: As you camp more and more, it will become obvious that certain tools are more than a nice to have. Having attempted to pound more than one tent stake into the ground with a rock, a hammer becomes an obvious tool to add in your tent bag. A hatchet, having a hammer face on one edge and a tool to aid in chopping or cutting wood on the other serves the function of two tools that can prove invaluable. A small, large toothed bow saw is also very convenient for cutting firewood from fallen limbs and trees. A shovel is another must have tool. Even in developed campgrounds, it is often necessary to dig out established fire pits. In undeveloped camping areas, a shovel is a necessity for preparing and covering a latrine, if you do not have a portable toilet, as well as clearing and preparing a fire pit.

Another convenience is heavy duty rope or nylon cord. This can be used to strap things down or help hold a tarp in place.

Fire: Always adhere to the fire regulations and restrictions of the area you wish to camp. Even if there are no fire warnings or restrictions posted, common sense is invaluable. Most of the forest fires in recent times which have burned hundreds of thousands of acres can be attributed to man's cavalier attitude towards fire, cigarettes, and the environment. Building a fire in an area surrounded by dry grass, leaves, or pine needles on a windy day can be fatal. Fire can bring comfort, warmth, light, and a feeling of security but left unattended or started in the wrong environment, fire can be extremely deadly and costly. Some National and State parks will only permit fire at certain times of the year and/or in designated fire pits. There are time periods when the fire danger is so high that all fires are banned. Those individuals who smoke are required to smoke inside.

The area around your fire should be cleared of all flammable materials such as dried leaves, grass, branches, or brush. Fires should be built in a fire pit only. Before leaving, your fire should be doused with water, the embers stirred, and doused again. The fire pit should then be filled with the dirt previously removed from the pit. Although fire has not been previously addressed, it is important to be able to build and ignite fires. Fire serves as warmth in the cold, light in the dark of night, and a means to warm and cook food.

When camping, it is always good to have multiple ignition methods such as matches and lighters. Most remember the bright white light of the burning magnesium in the science lab in high school. Magnesium burns at a temperature of approximately 5,400 degrees Fahrenheit, more than sufficient to ignite materials. Magnesium fire starter blocks can be purchased at most sporting goods or outdoor stores as well as online through sporting outfitters. Although ignition stable in block form, small slivers of magnesium may be shaved from this block and placed into a small pile on paper or dry tender such as dry leaves, grass, or pine needles. By striking a knife blade against the flint on one side of the magnesium fire starter block, the spark can ignite the shavings, thus becoming the ignition source of a fire for the most stubborn of tender.

Cooking: This subject is typically avoided in a disaster situation for several reasons. In each disaster, there is always a report of someone dragging their barbecue grill inside either to cook or heat. The inevitable result is a report of one or more individuals suffering from carbon monoxide poisoning. Often, the entire family succumbs to this silent and deadly killer. People have even become overcome by carbon monoxide poisoning in a tent.

Carelessness with fire in the home following a disaster can also be deadly. Often, the firefighters are the first responders to the most devastated and critical need areas. A house fire may not be one of the most critical of needs in the community. In some disasters, such as earthquakes, there may be a loss of water pressure, limiting the firefighting capabilities of the fire department. In the event of a natural disaster, or even in a camping situation, cooking must occur in a well ventilated area and as with any fire; care must be taken to keep flammable materials away from open flame. With those issues addressed, a small two burner propane camping stove is a real plus in the cases where power is lost and your ability to heat and cook food is lost as well.

Shelter: Most survival plans are based on the assumption that your place of shelter remains in tact and you are just isolated for a few days awaiting relief. It is possible your home could be damaged, missing a roof, or unsafe to remain inside. In other cases, you may be required to abandon your home and flee. There is always the shelter of the vehicle in an evacuation environment. The convenience and ease to erect a four person dome tent cannot be overlooked. Most can be erected in minutes by a single person and will provide some protection against the elements. By simply rearranging some furniture in a house without a roof, a tent or tarp may be erected to protect against the rain or snow. The plastic sheeting or tarps identified in the disaster preparedness kit can also be used as additional shelter or to cover your disaster preparedness kit against the elements as well.

Cots and air mattresses: After a night of fitful sleep on the hard ground, the comfort of cots and air mattresses cannot be overstated. While some air mattresses generally require an air pump of some type or a manual effort by the user, there are others that are self inflating or can be inflated by a small hand pump. Even these thin insulators of air between your sleeping bag and the cold hard ground are a plus. Combined with a cot, you may find that you will be able to sleep through the night.

Lamps and Lanterns: These also are a great plus and convenience. There are any numbers of battery powered lamps and lanterns available as well as gas and propane powered lanterns. It is far more convenient to cook, play games, and function in a stressful or camping environment by the light of a lamp or lantern rather than flashlights or completely in the dark.

Coolers: Although not covered in the disaster preparedness list, these can serve multiple functions from the convenience of storing portions of the disaster preparedness kit to saving and storing some perishable foods in the event of loss of power. When possible, it is wise to keep two to three bags of either block ice or crushed ice on hand in the freezer. The one gallon jugs of potable water mentioned as optional in the previous listing could just as easily be frozen and remain frozen until they are needed. By doing this, they can be used to fulfill the option of having additional water beyond the bottled water but the frozen blocks may also be used in the coolers to keep items cold when power is lost. Ice cubes cool items quicker than ice blocks, but ice blocks will last in excess of five days when the cooler lid remains closed, even on hot summer days.

There are advocates for the large 100 quart coolers. They require a lot of ice to chill and can become extremely heavy. A larger number of smaller sized coolers may be used to store your disaster preparedness items. Their size and loaded weight would be such that one or two people could easily load or unload them from a storage area to a vehicle or from a vehicle to camp. When used as intended for keeping items cold, it takes less ice per cooler to chill smaller coolers and by segregating items in coolers such as meats in one, condiments and milk in another, the coolers will remain colder longer due to reduced opening and closing to retrieve items. When using coolers for food storage, the old adage *'when in doubt, throw it out'* must be followed. Consuming items such as meat, eggs, items made with eggs, mayonnaise, or items made with mayonnaise that has spoiled or been left out of a refrigerated environment for any length of time can result in food poisoning which has the potential of being fatal.

Food Handling: We are constantly warned through public broadcast messages and commercial advertising about the dangers of handling and co-contamination of raw foods as we prepare meals for our families. In camping or in the event of a disaster, this warning should be amplified. When camping, we often associate 'roughing it' with not bathing or washing our hands. In a disaster situation, because of water constraints, we also tend to not wash our hands as frequently as we should. However, in both examples, our potential for handling and preparing food with dirty and unwashed hands increases. Hands can be washed and rinsed with little water usage. The use of anti-bacterial soaps and lotions is absolutely a must in handling foods both before and after food preparation.

Wash Basin and Dish Pan: The convenience of a small shallow plastic container about 18 x 12 x 4 will serve well as both a basin to wash hands and face but also a convenient vessel to wash pots, pans, and any cooking utensils.

Other conveniences: Although not necessary for survival, the conveniences of folding chairs and folding tables cannot be overlooked. Rather than the cumbersome lawn chairs or the bulky plastic chairs that can be purchased at many lawn and garden or hardware chains, collapsible heavy cloth or canvas chairs can be purchased inexpensively. These normally have carrying cases which allow you to fold and easily transport the chair to another area.

Chapter Seven

"In a moment of decision the best thing you can do is the right thing. The worst thing you can do is nothing." —Theodore Roosevelt

Evacuation

As we have witnessed time and again, when an evacuation order is given, people hesitate at first and then leave in mass. There are long lines at the gasoline stations to top off with fuel. Soon the gas station is out of fuel. In mass, the evacuees appear to leave their homes. Some vehicles with luggage racks full of suitcases. Most cars carry a single person. The result is hundreds of miles of gridlock. Some evacuations have resulted in excess of twenty hour duration for a travel distance of two hundred miles. Gasoline stations along the way were out of gas or closed as station owners and their families also began to evacuate. Families were caught without food. Few rest facilities available. People drove without air conditioning with both the elderly and very young in their vehicles. Others were able to push their cars and still remain with the flow and speed of traffic in order to conserve gasoline. Some cars overheated and had to be pushed to the roadside and abandoned.

Just as there are emergency preparedness kits of food, water, medication, other essentials in the disaster preparedness kit at home, there should also be a vehicle evacuation preparedness kit. Some people use or travel with such a kit on long road trips. This kit needs to be ready and available in case of an evacuation just as you have prepared a disaster preparedness kit in your home.

Fire Extinguisher: Except for a car jack, this is probably one of the most important items you should have in your car. They are relatively inexpensive and invaluable when needed.

Car Jack, Lug Wrench, and Cheater bar: Flat tires have a way of happening at the most inopportune moments. Having a car jack, knowing where the pieces are stored, and knowing where to place the jack and how to use them is a must for all drivers. A driver should know how to properly chock an automobile's wheels to keep the vehicle from possibly rolling off the car jack. Knowing how to remove the lug nuts and having the physical upper body strength to execute may be two different cases. A cheater bar is usually longer than the lug nut handle and equipped with the proper size lug nut socket, can often overcome the lack of upper body strength. If you have locking lug nuts installed, you must know where the key is located and how to use the key to remove the locked lug nut. Many automobile operators today carry auto club memberships and would prefer not to get their hands dirty. In the case of an evacuation or disaster, the convenience of the auto club service may be lost.

A spare tire: Having a spare tire is only half of the issue. The spare tire needs to be verified as being to the correct tire and wheel for your vehicle. The correct tire pressure must be verified on a periodic basis. There is no difference in not having a spare tire than in having a spare tire that is flat.

Two cans of fix-a-flat: Fix-a-Flat is a brand name product of the Pennzoil Company and can be purchased in various sizes at most stores that carry automotive products. Often, this product can be used to refill a flat tire with sufficient pressure and a chemical sealant to prevent leaking; this will usually allow the vehicle to proceed down the road until tire repair service can be obtained. Be prepared to accept that tire repair service may not be available or cannot be found.

Optional, Tire Plug Kit and portable air compressor: For the more mechanically inclined, Tire Plug repair kits can often be used to repair flats while the tire is still mounted on the vehicle. Generally, the process is to properly chock and jack the vehicle, remove the offending object such as a nail or screw and insert and partially pull the adhesive plug into and back out of the offending puncture. Then with a portable air compressor which plugs into the cigarette or other power access inside the car, refill the tire to the recommended tire pressure.

Roadside Flares: Most police cruisers will carry twelve to twenty-four of these flares to help warn drivers or guide drivers around road hazards. The purpose is to help protect lives that could be lost as the result of an oncoming driver not seeing the hazard in time to avoid an accident. Police cruisers may not always be available on a rainy moonless night when you find yourself on a darkened road, off to the side with a flat tire. Flares are inexpensive, and will last a long time in storage until needed. Carrying a dozen flares for your own emergency or to help a fellow motorist without flares, takes little space and can be a true life saver.

Fluids: There are four containers that should always be in your roadside emergency kit; a quart of oil, a quart of transmission fluid, a gallon of anti-freeze, and a gallon of water.

Flashlights: It is recommended that a flashlight be carried at all times in the glove box or center console of the vehicle. In the emergency roadside kit, an adjustable 6v waterproof flashlight is recommended. By being able to sit the base of the light on the road and adjusting the beam on the tire and wheel, even on the darkest night, enough hands free light can be focused on the task of automotive repair. As with any battery powered device, batteries must be checked periodically to verify they still work and have not corroded and damaged the flashlights.

Glass cleaner and paper towels: The convenience of a small bottle of window glass cleaner and a roll of paper towels cannot be overstated. The cleanser can not only be used to clean glass, but may also be used to remove the bulk of grim and grease acquired from changing a tire.

Battery jumper cables: This is another item that should be kept in vehicles at all times. Whether it is a cold winter day in Minnesota, a hot summer day in Arizona, or just the day you forgot and left your lights on, you will need a battery jump someday with your car. Having battery cables and knowing how to connect and use them are a plus. As previously stated many automobile operators today carry auto club memberships and would prefer not to get their hands dirty. In the case of an evacuation or disaster, the convenience of the auto club service may be lost.

Extra fuel: It would be nice if there was always a service station with gas available just before we ran out of gas. It is a sound practice to never let the gasoline gauge in your automobiles or trucks drop below ½ full. That usually means you have slightly more than 1/3 tank of gasoline. If possible and practical, it is a good practice to keep approximately five gallons of gasoline properly stored and available for emergency use. Both of these practices combined, in a two car family, insures that there is sufficient gasoline to provide a full tank of gasoline for the initial leg of any evacuation plan. By using a gasoline siphon pump, (approximately $10.00 in cost) gasoline can be safely transferred from one car to the next. Combined with the five gallon gasoline reserve you will be able to avoid the long lines waiting at the gasoline pumps. In the event of an evacuation and the emergency reserve is not consumed, it should be secured and transported safely until needed. It should be recognized that trying to hand carry five gallons of gasoline anywhere of any distance is next to impossible. (The gasoline will weigh in excess of forty pounds.) For transport to a car without fuel, a one to two gallon gasoline container is suggested.

First Aid Kit: Although it is not as necessary to carry as complete a First Aid Kit as described in the disaster preparedness kit a quality First Aid Kit should be carried at all times in your vehicle.

Maps: Carrying maps for the local city, state, and region, is a best practice. A warning of a bad accident, train derailment, flooding, or the collapse of a roadway may require you to detour from your normal route. Although emergency authorities may be on scene and able to provide a detour around the problem, there may be times when you must be self reliant. Maps are inexpensive and take little room in the glove box or center console.

Other tools: The convenience of a multi-function utility tool in the center console or glove box cannot be overlooked. If you have a gasoline siphon pump, include this in your disaster kit for your vehicle as well. A tow strap can also prove to be a life saver in either pulling someone else out of a difficult spot or in pulling your own vehicle out of a ditch or snow bank.

Preparation for a winter travel: Winter possesses its own set of unique conditions. We take travel in any weather for granted. We assume in our travels that our journey from point A to point B will be uneventful and that we will reach our destination and return to warmth and security outside of our vehicle in a modest time table. If trouble occurs, we can always flag someone down or reach someone on our cell phone. Yet each year there are reports of travelers being stranded on roadways due to inclement weather. Invariably someone is hospitalized due to frost bite and one or more succumb to the harshness of being stranded in winter either through hypothermia or carbon monoxide poisoning.

Beyond the items listed previously, snow chains, sand or kitty litter, rock salt, and shovel may be desired tools to have as well. Travel with the proper attire. Even in a well heated vehicle, traveling in a simple sweatshirt or sweater during winter is not wise or advisable. Engine failure for any reason suddenly thrusts you into the frigid environment when the warmth from the heater ceases. Although the vehicle may shelter you from the wind and the affects of wind chill, the inside air temperature of the vehicle will soon stabilize at the equivalent outside air temperature. Someone ill prepared for freezing or sub zero temperatures can succumb to hypothermia. Longer terms of exposure can result in tissue freezing and beginning to decay while life remains in you. Such are the results of frostbite.

Always travel with adequate clothing including coats, gloves, heavy socks, and boots depending on the expected conditions. Should you become stranded and unable to reach alternative shelter, a heavy blanket or an appropriately rated sleeping bag will help you to survive. A wool blanket, even when wet, will still provide insulation and warmth. It is also well advised to travel with potable water and a few emergency provisions such as protein bars.

More than a good practice, advising someone of route, expected arrival and/or expected return can be the one thing that ultimately saves your life. This practice should not just be followed in the winter or during inclement weather, but is a practice to be followed on any trip of duration.

In the case of an evacuation or a road trip of some expected length for any reason, the vehicle emergency preparedness kit should be one of the last items loaded. The fire extinguisher, jack, and tire tools should the most accessible items after the vehicle is loaded. In winter, the location and availability of a shovel, sand or kitty litter, rock salt as well as snow chains should also be easily accessible.

As a head of the household, there is no need to await an evacuation order. In a terrorist attack, it is possible that the normal means of receiving official information has been disrupted. When either an evacuation order is issued and communicated by officials or if it is in the interest of our family's safety and security, we should be prepared to evacuate within two hours. In order to do that, you must have planned exactly what to take. You must have options for evacuation routes. In the case of floods, fires, or earthquakes there is a possibility of roads being impassable either due to flood waters, fires, or collapsed bridges. Depending on the nature and type of terrorist attack, preparation must have taken place to leave via major highways, city streets, alleys and back roads. If the attack is biological or gaseous and airborne, having already developed options for directions to leave, it is not difficult to make the decision to go.

Maps are an essential part of your flight package. City, state, and area maps such as Southeastern United States should be in the vehicle. Depending on your depth of preparedness, having copies of USGS topographical maps can also be a plus. In certain areas of the Southwestern United States, those maps provide Forest Service and other access routes that are not on the normal state map. These can be very beneficial in being able to avoid large traffic flows or congestion and by taking the Forest Service roads; you may be able to merge back to your original escape route having avoided traffic congestion.

In an evacuation, people tend to congregate on the Interstate highway system. That system was first started under President Eisenhower. As a young army officer in the early 1900's, the future President of the United States led a convoy of army vehicles across the United States. It was a grueling trip in early versions of trucks on roads that more closely matched the ruts created in mud of horse drawn wagons and carts. It was the remembrance of that journey coupled with seeing the autobahn in Germany during World War II that motivated President Eisenhower to push for an interstate system in the United States. It was part of the strategic plan in the event of a nuclear attack. Through out the interstate system there are sections that run for more than a mile as straight and level. These sections of the interstate system were established to enable military aircraft, if necessary, to utilize the interstate as an emergency landing strip.

The system was continued to be built through the sixties and for the most part was completed in the nineteen seventies. Traffic has more than quadrupled from the time of original design. The gross weight of freight trucks has doubled in most instances with a corresponding increase in the number of wheels dispersing that load. The system is in need of repair in some states with too few lanes for safe travel at the posted speeds when combined with the high volume of freight trucks. In some states there seems to be continual cycles of construction and repair to the interstate system. Most of the old U.S. Highway system still remains passable and only in a few instances the interstate and the old U. S. Highway route has become the same. While the old U. S. Highways may have curves and hills not found on the interstate system and speed limits normally posted lower, they were the primary road system for the Untied States prior to the interstate system. They are generally in a state of good repair and today are primarily used by automobile and local traffic. Given that the primary traffic flow will remain the interstate system if passable, the U.S. highway system may be a good alternative route of escape.

The Global Positioning System (GPS) was originally a military tool. There are many commercial hand held devices as well as some vehicles which are equipped with GPS and mapping software. The hand held devices are rather inexpensive and are excellent for locating ones self on a USGS topographical map. Many devices can be load from your home computer with maps as well. Not knowing what situation you may find yourself or your family, this is a good investment when using topographical maps or just determining if you are headed North, South, East, or West.

You must be prepared to take only those things absolutely necessary. Packing all the closet contents into suitcases, takes time and is impractical. Grandma's rocker is a wonderful object of great sentimental value but it remains a material thing. Although granny's rocker truly could not be replaced because of the sentimental value, your survival and the survival of your family are your objectives and take precedence over all else. Take only what is necessary for the family. Load your disaster preparedness kit. Use a checklist to make sure everything has been included. If all the items remain centrally stored and intact, except grabbing the documentation and identify folders as well as cash, all other items will be in one area. If you have time and are evacuating in front of a forecasted disaster, there may be time to save the family album and the pictures on the wall. But in these uncertain times, you must be prepared to walk away taking only the essential documentation and a hope that someday you may be able to return.

Be prepared to defend your life and your family's lives. If it becomes common knowledge that you have established an emergency reserve of cash, there may be those who will attempt home invasions, burglary, or strong armed attacks. There is no one more vulnerable than a person with a gun and unwilling or unable to use it. If you arm yourself in order to defend yourself and your family from attack, learn to shoot. Make certain your wife or significant other understands the gun and its safety features. Learn how to keep and store a gun securely in your home as well. Gun locks, gun safes, and bionic keyed lock boxes only work if they are used every time. Children can be very curious and determined in a quest. To them, people get shot all the time in movies and they are either up walking around as if nothing happened within minutes or they appear in another movie on a different channel the same day. Even cartoon characters return to the living, the same vibrant creatures they were. Dead are revived in video games by a simple restart. Death and the permanency of death is not a concept understood by children. Regardless if it is the pro-gun lobby or the anti-gun lobby, both agree there were over 100 children in 2004 that died as a result of gun related accidents. Learn and understand when deadly force can be used. As a general rule, fear for your life or the life of another is the criterion for the use of deadly force. Also understand that in an armed confrontation with an intruder, the loss of control of your weapon usually results in your injury or death. Learn the laws of your specific state. Know if your gun must be registered and if there are safety and shooting courses you can take locally. Some states recognize other state gun ownership and registration while others do not. The same applies to concealed carry permits. Understand that if a pistol is fired indoors, the bullet may pass through several walls, both inner and outer, before its energy is spent. Remember, your children may be lying just through the adjacent wall. There is ammunition available that can still be used for self defense but does not have the power to penetrate and travel through partitions and outer walls. The ammunition retains its lethality. It is a similar product carried by the armed sky marshals that travel on domestic flights.

Be prepared to travel light. Wear a good pair of comfortable shoes in case you have to walk some distance. Wheeled, carry-on-luggage is very convenient on smooth flat surfaces and for air travel but is not designed to be pulled over uneven surfaces. A small backpack for each person should be prepared containing a single change of clothes such as jeans, clean socks, and underwear. Put a solar blanket in each backpack. Have matches in a waterproof container or a reliable cigarette lighter, as well as aids to start a camp fire if needed. Have a First Aid kit. It is better to never need it than need it and not have it. Carry a bottle of aspirin. Aspirin still remains the number one drug for aches, pains, and anti-inflammation. If you find you have to leave your vehicle, you may wish to take the First Aid kit, flashlights, batteries, and toilet tissue from the emergency preparedness kit in the car. Also in each backpack a small flashlight as well as a small roll of toilet tissue in a zip lock or other water proof container is a must. A couple of Power Bars and water should also be carried in each pack. Available today are backpacks with "camel pack" water containers built in allowing you to take a drink as needed without stopping. Smaller children may not be able to walk or carry a backpack and may have to be carried themselves by their parents. Front and back carriers for small children are available. There is any number of small, battery powered, hand-held, radios for two-way communication that are very convenient should family members become separated while hiking or camping.

In cases where an evacuation must occur during the winter months or when nights are cold, individual sleeping bags should also be loaded into the vehicle. Sleeping bags are rated by their ability to keep a person warm within specific outside air temperature ranges. Although it can feel cramped and uncomfortable sleeping inside the car, in a sleeping bag insulated from the outside elements, attempting to sleep inside the car may be the best option. Periodically running the car's heater for warmth can be extremely dangerous. Not only does this action consume gasoline which may be a precious commodity, conditions then exist for the risk of carbon monoxide poisoning inside the vehicle. If you awake before this odorless, colorless gas, becomes fatal, you will awake with a flushed face and normally a massive headache. You need fresh air and immediate medical attention.

In America today, the work week typically finds the family scattered. Children are in daycares or schools. Husbands and wives work two jobs often across town from each other. In the event of a terrorist attack, a chemical spill as the result of a train derailment, or a fire that prevents the family from returning home, the family must have a plan. It may not always be safe or even possible for the family to meet at the house. Which parent retrieves which child or children must be thought out. If communication service is not disrupted it may be overloaded rendering the idea of "we'll talk if it happens" ineffective as a plan. The family should plan meeting place contingencies outside or around the city depending on the type of problem. Perhaps it is a truck stop, park, or parking lot. It could be at a friend's house, a church, or a mall. If possible, someone outside the immediate area should be designated as a contact point to coordinate with in order to connect with family members. A primary and an alternate contact are recommended.

Get to know your neighbors or those with whom you may attend a worship service. Learn where people work and have their telephone numbers and share your numbers as well. In many neighborhoods, the closest we come to meeting our neighbors is to wave at them as we drive by or we may speak to them in the elevator. Neighborhood watch programs not only help to protect and warn each other when strangers or persons who may have criminal intent come to our neighborhoods, it also requires neighbors to meet and talk. Telephone numbers and names are shared. Friendships may actually develop. We may find there are common interests. Neighbors may be willing to look after your pets and you theirs in some temporary situations. We may learn that there are neighbors who are home during the day that could contact us in the event of an emergency at our house while we are away or they may be someone we could contact. Should an evacuation become necessary for any reason, rather than have two vehicles on the road with one or two people each, pool your resources and evacuate together. If there is a neighbor or fellow worshipper who does not have a means of escape, share space in your vehicle.

Useful Web Addresses

www.ready.gov Provided by the Department of Homeland Security and provides threat status and levels as well as providing useful pointers and tips for being prepared for disaster.

www.redcross.org Provided by the American Red Cross. This site provides the capability to identify shelters and services as well as preparedness tips and checklists as well as links to other sites.

http://palimpsest.stanford.edu/bytopic/disasters/ This link is from Stanford University with links to a number of papers regarding disaster preparedness as well as post disaster recovery.

http://www.fema.gov/areyouready/ This link is provided by FEMA. The document, **Are You Ready?** Is FEMA's presentation of why you should be prepared and what is needed to be in your disaster preparedness kit.

www.prepare.org Provided by the American Red Cross. This provides the why, and what associated with disaster preparedness as viewed by the American Red Cross.

http://www.disasters.org/dera/weblink.htm This link is provided through DERA (Disaster Preparedness and Emergency Response Association, International) with additional web links for the Humane Society, Salvation Army, DERA and others.

There are also a number of sites that offer to provide a personalized emergency preparedness kit for your home, car or business for a fee. There are also websites that offer to sell and even install fallout shelters and bunkers. There are any number of books, pamphlets, and placards advising that we should all be prepared for disaster. In the final analysis, becoming prepared requires an acknowledgement that we must take responsibility for our survival. This may require that we review the assumptions base on a belief system that there is a safety net that will break our fall when disaster strikes. We must accept that we are responsible for the preservation and security of ourselves and our families. We are a nation willing to open our wallets and our homes to victims when disaster strikes. But the mobilization and transportation time for aid to reach our family should we become victims will not be instantaneous. We must be prepared to survive without outside support for a minimum of three days. It is a personal decision to be prudent rather than the simpleton. Preparedness is the most important decision that we will regret not having correctly chosen the day after disaster strikes again.

Printed in the United States
55490LVS00006B/169-216